ALLEN COUNTY PUBLIC LIBRARY
3 1833 05177 4062

Y0-BVQ-650

INVASIVE SPECIES

Invasive Microbes

INVASIVE SPECIES

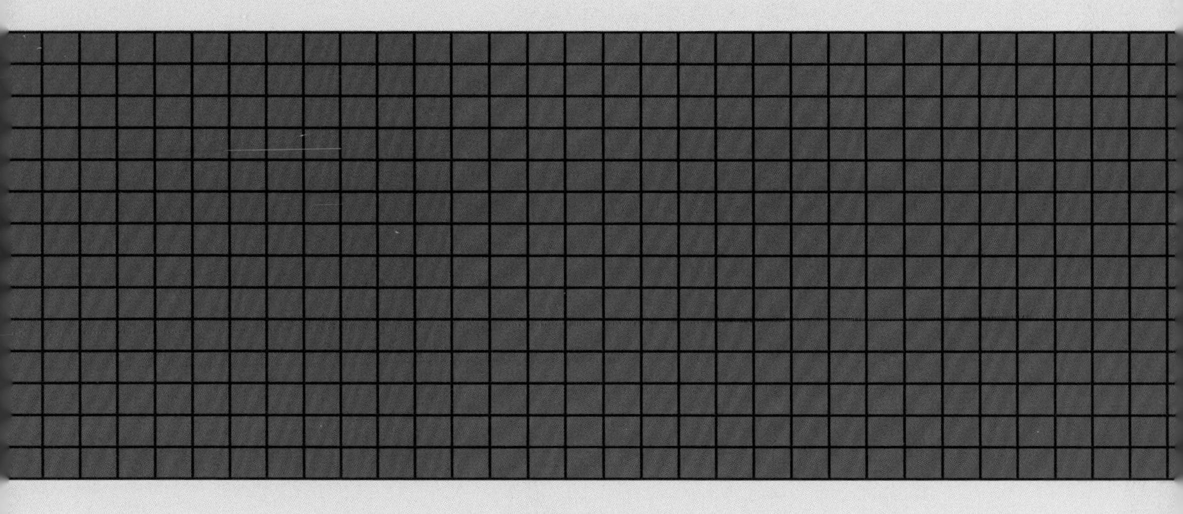

Invasive Aquatic and Wetland Animals
Invasive Aquatic and Wetland Plants
Invasive Microbes
Invasive Terrestrial Animals
Invasive Terrestrial Plants

INVASIVE SPECIES

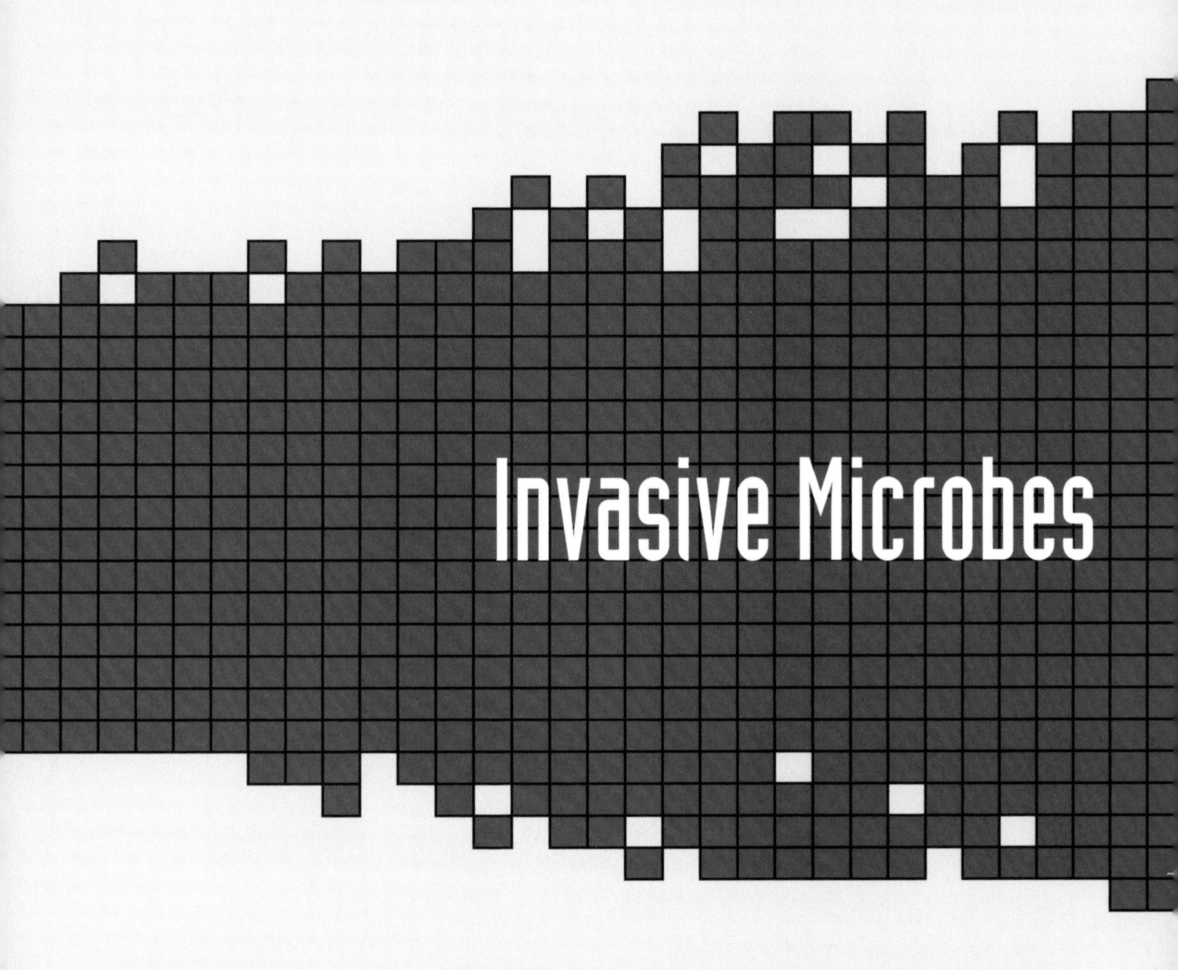

Invasive Microbes

Suellen May

Invasive Microbes

Chelsea House
An imprint of Infobase Publishing
132 West 31st Street
New York NY 10001

Library of Congress Cataloging-in-Publication Data

May, Suellen.
Invasive microbes / Suellen May.
p. cm. — (Invasive species)
Includes bibliographical references and index.
ISBN 0-7910-9131-7 (hardcover)
1. Microorganisms—Juvenile literature. 2. Bacteria—Juvenile literature. I. Title. II. Series: May, Suellen. Invasive species.
QR37.M38 2006
581.7'6--dc22
2006011031

Chelsea House books are available at special discounts when purchased in bulk quantities for businesses, associations, institutions, or sales promotions. Please call our Special Sales Department in New York at (212) 967-8800 or (800) 322-8755.

You can find Chelsea House on the World Wide Web at http://www.chelseahouse.com

Text design by James Scotto-Lavino

Cover design by Takeshi Takahashi

Printed in the United States of America

Bang EJB 10 9 8 7 6 5 4 3 2 1

This book is printed on acid-free paper.

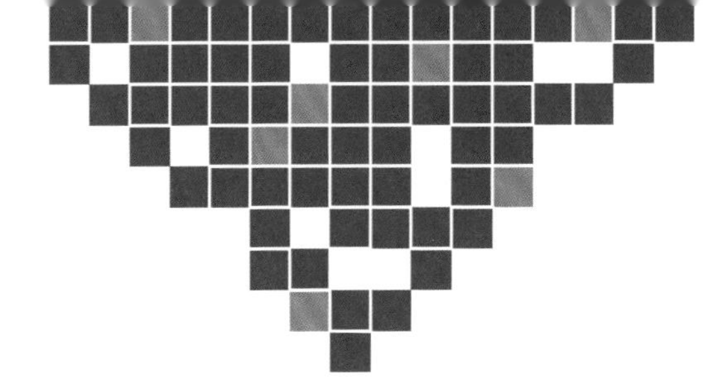

TABLE OF CONTENTS

1 Microbes

MORE THAN JUST GERMS

Microbes are living creatures that are so small they cannot be seen with the naked eye. You need a microscope to see them. In some ways this might be a good thing. Can you imagine what it would be like to see all the bacteria and fungi living all over your body?

More microbes exist on your hands than there are people in the entire world. But it does not stop at your hands. At this very moment, in the cubic yard of air at the tip of your nose, hundreds of thousands of microscopic bacteria, viruses, fungal spores, algae, and pollen grains are floating by.[1]

Microbes consist of cells just as humans do. When examining the entire human body, it turns out that less than 10% are human cells (Figure 1.1). The remaining 90% are viruses, bacteria, fungi, protozoa, worms, and insects.[2] Most of these organisms do us no harm.

Microbes are just doing what all living things do: looking for food, excreting waste, reproducing, forming communities, and evolving. The goal of any living creature is to survive and reproduce. Whether it is an insect, a snake, or a human, all living organisms strive for these goals, even if it means negatively influencing other living creatures. Microbes are no different.

One major difference between microbes and other living creatures is that microbes are so small they function without a circulatory system. The human body is much larger and has

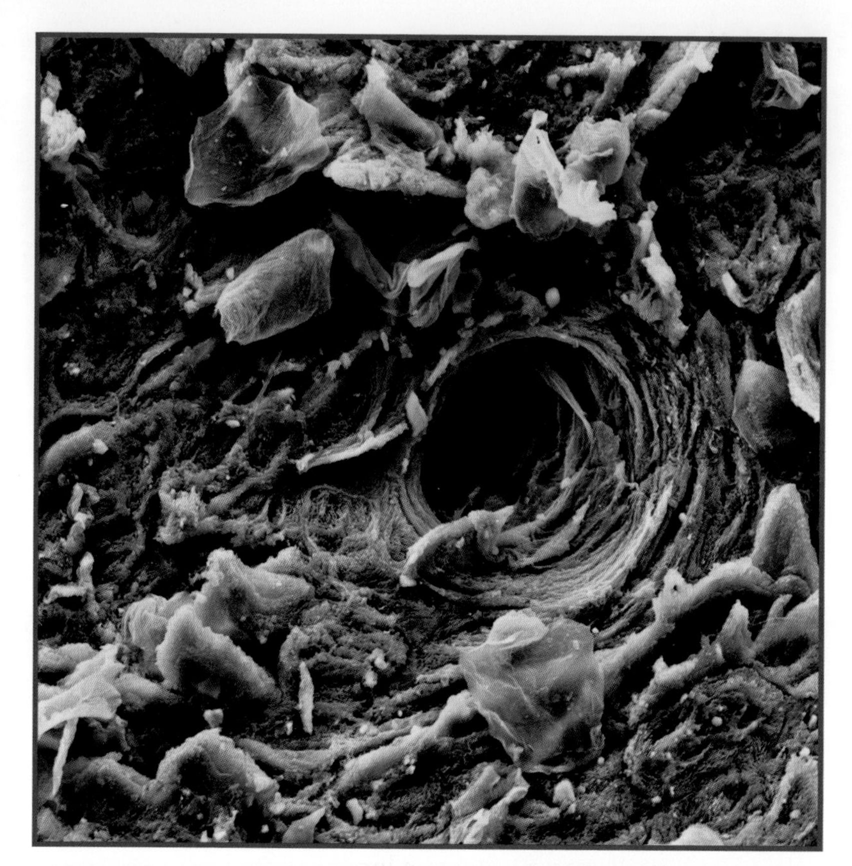

Figure 1.1 The human body is covered in microbes that luckily cannot be seen with the naked eye. With a microscope, skin would appear as pictured in this photograph. This microbe, *Staphylococcus*, is a bacteria that forms colonies on human skin and can cause boils or infection.

a circulatory system, which delivers oxygen and nutrients to tissues and gets rid of waste products. The circulatory system uses the heart and blood vessels, which move blood throughout the body, to accomplish this. Fish need gills to expand surface area and access the dissolved oxygen in the water. Microbes have no circulatory system, yet they still prosper.

Microbes have been around longer than any other group of organisms. If you imagine all time since the Earth began as a single day, microbes would have appeared sometime about 5:00 a.m.; dinosaurs do not pop up until about 10:00 p.m.; and humans do not appear on the scene until seconds before midnight.[3]

Microbes may be small, but they are able to do remarkable things. They can move through air and water, resist others, communicate, feed, learn, and adapt to their environment. Some microbes move around with cilia or tiny hairs; others use an oarlike appendage. Stationary microbes modify their environment by secreting chemicals to draw other microbes near instead of putting energy into moving.

WHERE THEY LIVE

Microbes have a place of origin just as animals and plants do. The kangaroo is native to Australia; the aspen tree is native to the United States. Animals and plants can live and thrive in other countries, but they have one specific place of origin. The native country or region is where the organism changes over time in response to its environment. A microbe may become more competitive in a new location because predators of that organism do not exist in the new location.

Microbes Live on Bodies

A body makes a lovely home for a microbe. Humans and animals are home to an amazing array of microbes. The idea may be unnerving, but in most cases the microbes do no harm. It is not in an organism's best interest to debilitate or kill its host.

Being the host for microbes might not be fatal, but it can be unpleasant. Athlete's foot is the result of fungi taking up residence in moist spaces, particularly between toes. This condition occurs when a particular fungus rapidly increases its population. Although it is called athlete's foot, this fungus can also grow on the hands; this is less likely, however, as hands are not usually as warm and covered up as the feet. Athlete's foot is contagious and must be treated with an antifungal medicine.

Not all host and microbe relationships are bothersome. Some can be relatively harmless or even beneficial to the host. When a microbe is living in a state of harmony within the body of another organism, this is known as **endosymbiosis.** An endosymbiotic relationship is one in which both the host and parasite benefit. Another way to describe this relationship is **mutually beneficial.**

Termites have an endosymbiotic or mutually beneficial relationship with microbes. Surprisingly, termites are not able to break down the main ingredient in wood and dead plants without help. Termites enlist the help of microorganisms, such as bacteria and protozoans, to digest cellulose. These microorganisms live in the hindgut of the termite's digestive system. The microorganisms benefit by getting a cozy home with food delivered directly to them. Termites would literally starve to death without these organisms, even if they had a steady supply of wood. The microorganisms are equally dependent on the termites for their survival.

Humans also need microbes. Similar to the termites, humans need microbes to inhabit our body for our own survival. The minute we are born, we leave the sterile environment of the womb and the microbes move in. Microbes in our intestines make vitamins to support the human body. *Escherichia coli* bacteria are like little factories making and exporting vitamin K and some B vitamins. If you are not getting enough protein in your diet, *Klebsiella* bacteria can provide your cells with raw chemicals needed to make protein.[4]

Microbes in Food

In 1993, more than 700 people got sick from eating undercooked fast-food hamburgers. These people suffered from bloody diarrhea and the potential threat of kidney failure. *Escherichia coli* was the culprit.

Escherichia coli O157: H7, as it is more formally known, is one subcategory of bacteria within the larger category of *Escherichia coli* bacterium. Most strains are harmless and live in the intestines of healthy humans and animals. This particular strain of *E. coli* produces a powerful toxin that causes illness. In people with fragile immune systems, including children younger than five, the infection can also cause a complication in which red blood cells are destroyed and the kidneys fail. This condition is called hemolytic uremic syndrome and occurs in 2 to 7% of all *E. coli* infections.

This microbe is found on cattle farms. When a cow is slaughtered, the *E. coli* microbes can be mixed into the beef when it is ground up. Most infections from *E. coli* have come from undercooked meat, but *E. coli* can also be found in unpasteurized apple juice, salad vegetables, yogurt, and drinking water. Bacteria on the cow's udders or on milking equipment can get into the raw milk. Pasteurization of milk and cooking of meat will kill the bacteria. These bacteria are also passed through to the stools of the infected animal and can be the source of infection to another person. Most people recover after five to ten days.

Microbes in the Soil

Scientists believe that, pound for pound, there may be just as much life living below the ground as there is life on the planet's surface. Tiny spaces in the soil provide oxygen for these underground microbes. Microbes that exist deep below the Earth's surface must rely on nearly no oxygen and little available food. Sustenance is so meager for these microbes that it may take thousands of years for these microorganisms to reproduce by dividing in two. This method of reproduction is known as **asexual reproduction.**

Plants and humans are fortunate that microbes can withstand these harsh conditions and still reproduce. Plants need these microbes to get the nutrients far below the surface of the ground. Mycorrhizal fungi are considered one of the most crucial types of microbes for plants (Figure 1.2). Mycorrhizal fungi live in and around plant roots and extend far out into the soil. These fungi get minerals and water out of the soil, which is then made available for the plant. The plants are not the only ones that benefit; the microbes benefit by keeping the plants alive because the microbes feed off sugars that the plant produces.

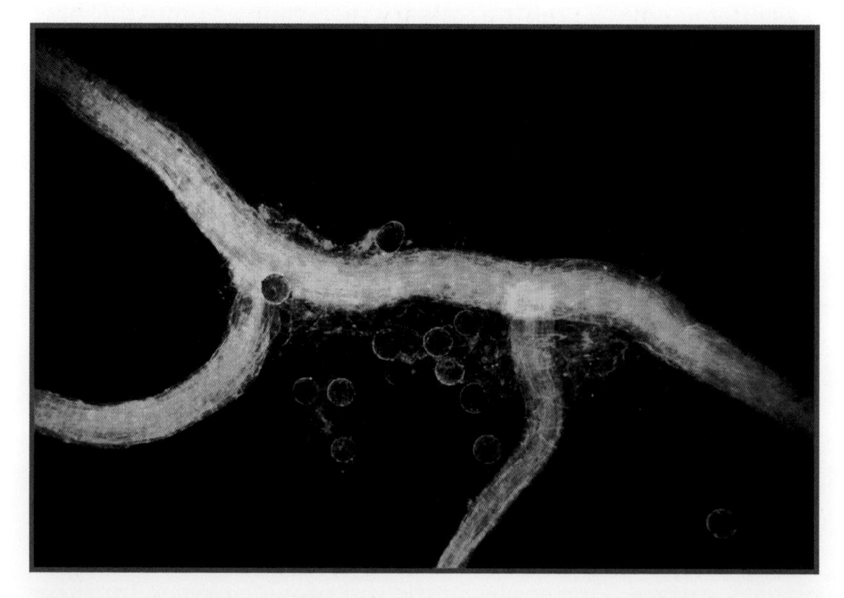

Figure 1.2 Mycorrhizal fungi are beneficial to plants. These fungi grow on plant roots. The round bodies are spores from the fungus, and the threadlike strands are the hyphae growing on a corn root. The fungi extract minerals and water from the soil and make it available to the plant. In return, mycorrhizal fungi live off the sugars that the plant produces.

Bacteria are also important in helping plants get nitrogen. Plants need nitrogen in large quantities to grow. Most fertilizers include nitrogen as a primary component. Nitrogen exists in the air, but plants are not able to access it. Some bacteria are referred to as **nitrogen-fixing bacteria**, because they "fix" nitrogen—change nitrogen into a usable form—to make it accessible to plants. The plants in turn accommodate the bacteria by creating root nodules in which bacteria live.

Microbes benefit crop plants by protecting them. Recently, scientists in Maryland were able to isolate a root- and seed-dwelling bacterium that protects plants from diseases. This bacterium protects plants from harmful microbes by producing an antibiotic. It is not clear why the bacterium produces an antibiotic, but it is common for living organisms to produce chemicals that are not necessary for their survival. These compounds, known as secondary compounds, include the antibiotic that this bacterium produces.

Bacteria can also increase the growth of plants. According to researchers at McGill University in Montreal, Canada, bacteria use chemical signals created by plants to stimulate plant growth. By increasing plant growth, bacteria increase the food supply that they require, and microbe and plant benefit. A young soybean plant releases chemical signals to attract the bacteria. The bacteria reply by sending their own chemical signals to tell the plant to make the root nodules where the bacteria live.

TYPES OF MICROBES

Bacteria

The majority of microbes on or in our bodies are bacteria. Feces contain a large number of bacteria. It is estimated that there are approximately one million million (10^{12}) bacteria per gram of feces.[5] Bacteria in its usual location or "colonization site," such as the gastrointestinal tract, do not pose a risk to your health.

Bacteria can reproduce in as little as 20 minutes. Studies show that this creates a high number of mutant bacteria. Nearly all populations of living things will have individuals with different genetic makeup, otherwise known as a mutation. These mutations are believed to occur by chance. Bacteria are no different and many are in a better position to adapt to their environment because they reproduce so quickly.

The mutation rate for bacteria is, at the lowest, 1 in every 100 million cells. This may sound like a small amount but, given the amount of bacteria present in the human body, that turns out to be a high number. If there are 100 million million (10^{12}) bacteria present in the intestines, there will be about 1 million mutant bacteria.[6] This means that the bacteria in the intestines would

Antibacterial Agents: How Are We Changing Microbes

Respiratory infections are the leading cause of death by infectious disease worldwide. Respiratory infections include pneumonia, influenza, and whooping cough. Respiratory infections are caused by microbes, specifically fungi, bacteria, and viruses with approximately 1 billion cases per year and a resulting 4.7 million deaths per year. More than 3 million people die each year worldwide from bacteria, viruses, and protozoa that give them diarrhea, such as amoebic dysentery. The remaining most deadly infectious diseases are tuberculosis (caused by a bacterium), malaria (sporozoan), AIDS (virus), measles (virus), hepatitis B (virus), and tetanus (bacterium).

have a great capacity to adapt to their environment as it changed. If antibiotics were taken that would ordinarily kill bacteria, mutant bacteria that have a resistance would further multiply creating an antibiotic-resistant strain of intestinal bacteria.

It may be unappealing to think of so many bacteria on and in your body, but a lack of normal bacterial populations poses a greater problem. In fact, burn victims suffer from infection because the skin no longer provides a habitat for beneficial microbes. The wrong kind of bacteria also poses a threat to your health. During surgery, bacteria can be inadvertently moved from one part of the body to the other to the detriment of the patient. An operation on the colon could cause the bacteria found there to spill into another part of the body and cause abscesses or areas of pus and inflammation.

Legionnaire's Disease

In 1976, members of the American Legion attended a conference at a prestigious hotel in Philadelphia. Many of the attendees developed a high fever, chills, and a cough. Without knowing it, these people had breathed in mist contaminated with bacteria. It appeared that they had pneumonia, but the bacteria that caused their illness had never been seen before. The symptoms were similar to many other forms of pneumonia so it was hard to diagnose.

Scientists discovered that the American Legion conference attendees were infected with a newly discovered bacteria, subsequently named *Legionella* in relation to the event. The disease is now referred to as Legionnaire's disease. To determine that a person has Legionnaire's disease, doctors do chest X-rays along with bacterial tests of their phlegm or mucus.

Archaea

Originally scientists thought of archaea as a weird kind of bacteria that live under extreme conditions: boiling temperatures and sulfur-spewing volcanic vents. Some archaea can tolerate temperatures as high as 239° Fahrenheit and live without sunlight. These kinds of archaea are often called extremophiles because they are able to survive extreme conditions.

Now, scientists know enough about archaea to put them in a different taxonomic group. As often happens in science, organisms are lumped together because we have not yet learned enough about them to understand their differences. When a group of organisms is understood well enough, it gets its own name.

In the classification of living things, one important distinction is whether the organism's genetic material is enclosed in a central cellular compartment, the **nucleus**. Organisms that do not have this genetic material enclosed in a nucleus are classified as **prokaryotes**. Archaea are prokaryotes. Life-forms with a nucleus are **eukaryotes**. Think "eu" for nucleus, therefore "eu" for eukaryotes.

The distinction of an organism having cells with or without a nucleus is an important one. It is so significant that the five-kingdom system of classifying all living things is now being revised to show this difference. Since 1969, the scientific community has recognized five kingdoms: Monera (bacteria and blue-green algae), Protista, Fungi, Plantae, and Animalia. Prior to 1969, there were only two kingdoms: Plant and Animal.

The five-kingdom system is currently being revised to expand the number of kingdoms and put them under another category of superkingdoms referred to as domains. The domains are Bacteria, Archaea, and Eukarya. The reason for creating an additional taxonomic category of the domains is due to the discovery that unicellular organisms without a nucleus (Bacteria and Archaea domains) are far more varied than once thought.

Multicellar organisms having cells with a nucleus (Eukarya domain) are plants, animals, fungi, and protists.

Fungi

Fungi include mushrooms, mildew, and molds. Fungi are organisms that have some form of a vegetative body, usually produce spores, and lack chlorophyll. **Chlorophyll** is the green pigment that enables a plant to use sunlight and carbon dioxide to make food to support its structure and function. Because fungi do not have chlorophyll, they do not need light to make food. Fungi instead nourish themselves by decomposing organic matter, such as wood. You will find the vegetative body in places high in organic matter, such as below the surface of the ground. From a human perspective, fungi can be beneficial or harmful.

Fungi do not contain a true stem or root. They can range from a single cell to chains of cells miles in length. Mushrooms are an example of multicellular bunches of fungi. The mushrooms that you see are actually the fruit of the fungus colony living underground.

The vegetative body of a fungus can have many forms. Hyphae is one form that consists of microscopic threadlike filaments that branch out and feed the fungus. Long, threadlike strands of hyphae have that fuzzy appearance often associated with fungi, specifically mold. Because fungi do not have a root system, hyphae branch out to access nutrients and supply them to the fungus. **Plasmodia**, another type of fungal vegetative body, are amorphous, jellylike slime molds.

Yeasts, another type of fungi, are independent, single-celled creatures. Yeast, although existing independently, cluster together and can be seen without a microscope. Many of them lumped together appear as a white powdery coating, particularly on fruits and leaves. Yeasts are the most important microbes in the food industry. They are used to make alcoholic beverages, leaven bread, and make foods tasty and nutritious.

A fungus can form mushrooms or mold and, although these organisms are readily visible without a microscope, they are considered microbes because the individual organisms are not visable to the naked eye.

Fairy ring is the name of a common disease in lawns. Mushrooms appear in a circular pattern, but the cause of these mushrooms is a fungus that lives off decaying organic matter such as a tree stump that was buried during the landscaping process. The fungus that causes fairy ring consists of a network of **mycelium** (dense mold). Fairy ring is considered a turf grass disease because the mycelium layer prevents water from infiltrating the soil and reaching plant roots.

Fungi can be an adversary to humans by causing athlete's foot or it can be helpful by being the source of medicines, such as penicillin, which are based on the natural traits of fungi to compete against bacteria for nutrients and space.

Fungi prefer acidic environments. Fungi can grow on your body, in soil, in seawater, on plants and animals, and in your house. Like all living things, fungi reproduce and spread their progeny. Fungi have two ways to reproduce: One way is by extending their hyphae and another way is by producing spores. **Spores** are the reproductive element of simple organisms, such as protozoa, fungi, and nonflowering plants such as ferns.

The advantage of spores is that, similar to seeds, they can be spread by wind or rain, or they may stick to people or animals and be carried farther distances than hyphae can travel. A new population will begin growing far from the parent fungus.

Viruses

It starts with a rapid onset of fever, malaise, muscle pain, and severe headaches. Then nausea is followed up by vomiting.

Often there is bloody diarrhea. Within five to seven days, the sick person shows signs of bleeding into the skin and mucous membranes. Death usually occurs after six to nine days. The infected person has come into contact with blood and tissue from monkeys from Uganda. These are the effects of the filovirus, more commonly known as the Ebola virus.[7]

The Ebola virus did not originate in humans; it was probably a harmless virus in bats or small rodents. Tracking the origins of these outbreaks is difficult. All human Ebola virus outbreaks from 2001 to 2003 in the forest zone between Gabon and Republic of Congo resulted from handling infected wild animal carcasses. Once the virus enters the human body, it can be spread by bloody secretions from one human to another. The fatality rate in human cases ranges from 50 to 89%, depending on the viral subtype.

Without international travel, this microbe would be confined to spreading from one host to another through contact with an infected animal carcass, bloody secretions, possibly sexual intercourse, or eating an infected animal. This limits the geographic spread of the Ebola virus. Today, things are different. A doctor taking care of patients in Gabon became infected. It takes at least four days for the virus to incubate before symptoms appear. This doctor flew to Johannesburg, South Africa, and was then admitted to a hospital. He infected a nurse who later died. Modern air travel helps spread this illness.[8] **Viruses** such as the Ebola are the most primitive of all the microbes. They do not have a true cell or nucleus. Some scientists even debate whether they are a life-form.

Viruses must attach to a host cell to reproduce. A virus first attaches to a host cell by recognizing a particular shape of a structure on the cell. This structure on the cell is known as a receptor. If a cell does not have an appropriate receptor, the virus cannot attach to the cell, and the cell is unaffected (Figure 1.3).

If the virus is successful in attaching to the cell, it enters the cell. After gaining entry, the virus shuts off the host cell's normal functions and redirects the cell into making new copies of the virus. Each infected cell can release thousands of copies of the virus, which then go on to infect other cells.[9]

Protists

Protists include algae, amoebas, and protozoa. Protists are eukaryotic, meaning that their DNA is enclosed in a nucleus. Algae is a familiar term to most. Algae thrive in the sea and on land. Although single organisms are microscopic, algae can grow together in blooms large enough to see. A notable distinction of algae is that they are able to make energy from sunlight, just as plants do. Whereas many microbes get their nourishment from a host, algae are somewhat self-sufficient.

The parasite that causes whirling disease, which will be covered in Chapter 3, is an amoeba, or a type of protozoa. Amoebas usually surround their food and engulf it; some amoebas have mouths. Amoebas use their pseudopods, or false feet, to move around.

PATHOGENS: DISEASE-CAUSING MICROBES

Jeffrey was enjoying snorkeling with his family while on vacation in Hawaii in April 2004. The eight-year-old boy did not think much of it when his shoulder started to ache. The pain got worse, and his shoulder became red and swollen. Eventually the pain became so unbearable that Jeff's parents took him to the hospital; they watched as an infection raced through his body and threatened his life. The cause was a microbe, a pathogen known as methicillin-resistant *Staphylococcus aureus* or MRSA. MRSA is a microbe that can get through a break in the skin and

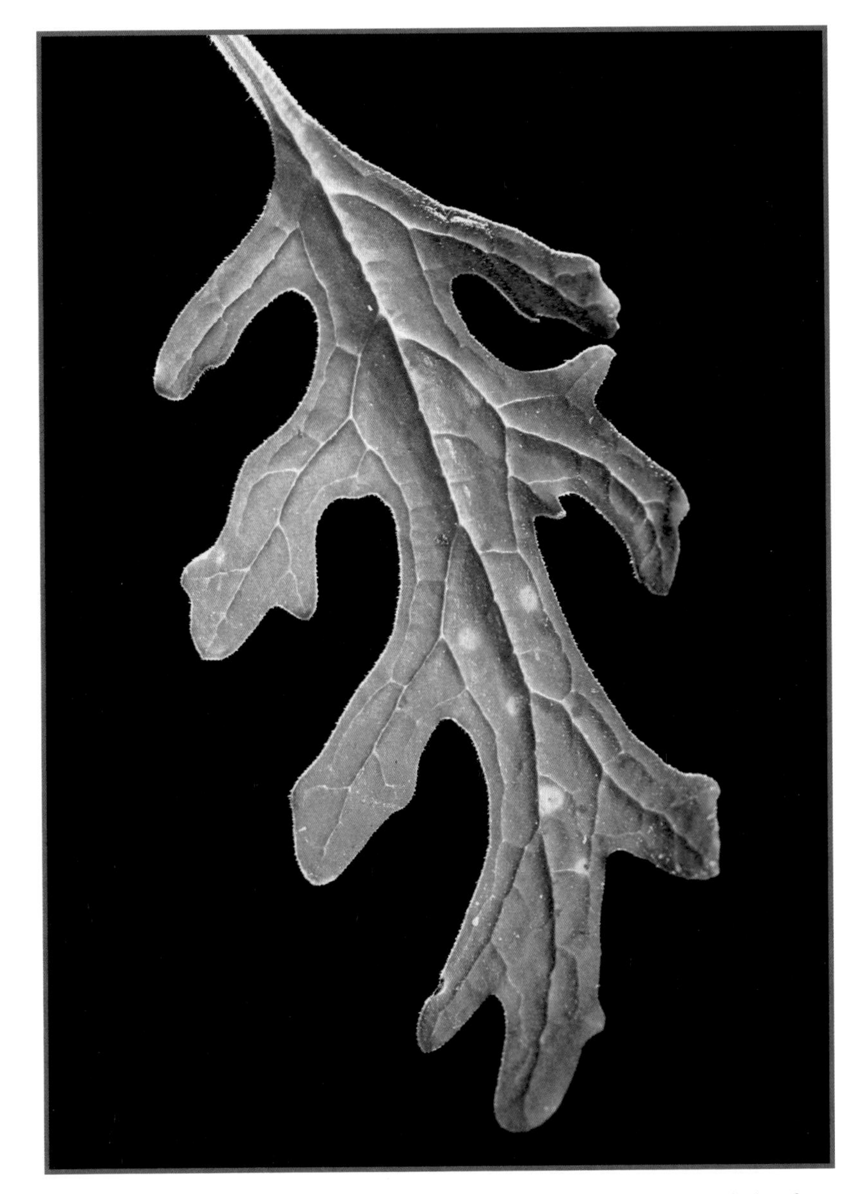

Figure 1.3 Once a virus has invaded a cell and replicated, it often shows signs of its presence, such as the yellow spots on these leaves. The plum poxvirus has infected this plant. A virus invades a host by finding a receptor on a cell and tricking the cell into making many copies of the virus.

cause an infection unless certain antibiotics are used to stop its spread. Unfortunately, MRSA is resistant to an antibiotic commonly used to treat this infection, methicillin, which makes it more difficult to treat. Antibiotics are not completely powerless against MRSA, but patients may require a much higher dose over a longer period, or the use of an alternative antibiotic to which the bug has less resistance.[10] Luckily, Jeff recovered by his ninth birthday.

Pathogens are disease-causing microbes. The virus that gives you the sniffles is a pathogen just as is the fungus that makes a crop rot. Microbes have greatly influenced human history. In William McNeill's *Plagues and Peoples*, he claims that infectious diseases have influenced the outcome of wars and shaped the location and development of human societies. Epidemics usually come from diseases that can be spread easily, such as through sneezing and other respiratory discharges. Smallpox and whooping cough are diseases that spread when coughing expels a tiny droplet of bodily fluid filled with the microbe that can cause the infection. Smallpox spreads through contact with skin lesions, linens, and clothing in close enough contact with the patient that bodily secretions are on them. The length of time it takes for the disease to incubate, or for a person to remain infectious, is also another factor contributing to a microbe's successful spread.

There is an important distinction between human-related and animal-related pathogens. Some pathogens have made the leap from animals to humans. The distinction is crucial to those who monitor and try to prevent global crises due to pathogens, such as the World Health Organization.

Scientists have counted 1,400 pathogens that affect humans.[11] When scientists look at human-related pathogens, they describe these pathogens as emerging or reemerging. An **emerging pathogen** is one that is relatively recent, such as

SARS (sudden acute respiratory syndrome) or HIV (human immunodeficiency virus). Reemerging pathogens are ones that were thought to be under control but are making a comeback, such as tuberculosis, West Nile virus, or malaria.

Humans have not lived on the Earth that long, and science has been practiced for an even shorter period of time, so it is difficult to tell why it seems that the number of pathogens is on

The Deadliest Infectious Diseases Worldwide

Antibacterial products are everywhere: soaps, lotions, wipes, and even sweatsocks. Washing hands with regular soap cleanses by washing off microbes. Antibacterial agents cleanse by killing microbes. If you take a careful look at the label of any antibacterial agent, you will notice that it shows the percentage of microbes that are killed. Although the percentage is often close to 100, some microbes do survive. The surviving microbes were resistant to the antibacterial agent and did not die.

Scientists have studied this effect and recognize that microbes in general have changed in response to the million or so pounds of antibacterial agents added to our environment each week. When bacteria is sprayed with an antibacterial agent, the resistant bacteria will continue to reproduce. These stronger, resistant bacteria will create a strain that cannot be killed with the antibacterial agent that killed the weaker bacteria. As a result of these concerns, some organizations suggest avoiding unnecessary antibacterial household products. Instead, simply washing hands for at least 20 seconds is sufficient to eliminate most of the harmful bacteria.

the rise. Scientists do know that human pathogens have emerged or reemerged hundreds of times in the past 50 years.

Most of these diseases are zoonoses. **Zoonoses** are diseases that come from animals. They are of serious concern because half of the emerging viruses are characterized as causing inflammation of the brain or serious neurological (nervous system) symptoms.[12]

REPRODUCTION

Humans reproduce sexually. Sexual reproduction occurs when two individuals create another individual with genetic makeup that is similar, yet different from the parents. Another way to reproduce is to make a copy of oneself by splitting in two, which is asexual reproduction.

The advantage of asexual reproduction is that only one organism is needed to reproduce; in sexual reproduction, an organism must find a mate. If a single microbe reaches a new habitat but cannot find a microbe of appropriate sex, it will continue to be a sole microbe without any offspring. Organisms that live in remote places where mates might be hard to come by often reproduce asexually.

The disadvantage to asexual reproduction is that a change in genetic material can only occur through mutation. In asexual reproduction, there is no incorporation of new combinations of genetic material. The offspring from asexual reproduction would have the same genetic makeup as the parent, unless a mutation occurred.

Sexually reproducing organisms have the advantage of more diversity in their offspring because a genetically different organism would be created. Traits that enable the organism to be more competitive would have a better chance of being passed on to successive progeny. Changes over time to make a more fit species are part of the evolutionary process. Sexually reproducing species do not have to wait for an organism to mutate for a new trait to occur that could offer a benefit to the species.

Most microbes are able to reproduce sexually and asexually, depending on the situation. A microbe might reproduce asexually when the microbial population is low, meaning that sexual partners are limited. If the population density increased enough, the microbe might then reproduce sexually. Scientists believe that microbes communicate through chemical signals to determine how many other microbes are present in their immediate environment.

Sexual reproduction involves eggs or spores that can remain dormant for a long time, in case the situation cannot support new life. The microbe is able to factor in environmental aspects, such as nutrient supply, acidity, and temperature extremes. The egg or spore is dormant and, when the environment becomes favorable, the microbial life cycle begins.

2 The Spread of Invasive Microbes

Invasive species include animals, plants, and microbes that infiltrate and invade ecosystems beyond their historic range. Their invasion threatens native ecosystems or commercial, agricultural, or recreational activities dependent on these ecosystems. They may even harm the health of humans. In some cases, the native species is completely displaced. Invasive species are generally not native; they are usually from another country or region. Other terms used to describe invasive species are invaders, nonnatives, exotics, invasives, and nuisance species. These terms will be used interchangeably.

Organisms in a new environment are not a new problem. Humans traveling across the Atlantic Ocean have assisted with this invasion for centuries. Many of these invasives entered the United States in the 1800s. In some cases, they were intentionally introduced, such as the brown trout (*Salmo trutta*), because benefit was perceived in their introduction. Many anglers wanted the brown trout here for sportfishing. In some cases, the imported species continues to make life easier. Many imported plants are an important source of food.

This travel transformed the existing plant and animal species. A hundred years ago, nobody realized the kind of ecological warfare these organisms would wage on native organisms, resulting in a smaller gene pool. Scientists have compared the movement of these exotic plants, animals, and microbes to a game of biological roulette. Once in a new environment, the organism might die or it might take hold and became an ecological bully by stealing nourishment and habitat from native species. Just like other newly introduced organisms, many microbes in a new environment do not survive. Many are not able to reproduce and therefore die out.

Today, we see the dwindling of native species caused by these invaders. Introductions are accelerating due to travel, trade, and tourism (Figure 2.1). Whereas oceans and unscalable mountains kept animals from entering terrain where they did not belong, trade and travel have bridged almost all geographic barriers. A plane can take you from Philadelphia to sub-Saharan Africa. If an animal finds its way into the cargo area of a plane, it could travel thousands of

Figure 2.1 International travel is just one of the ways invasive microbes have been able to move farther and faster around the globe. Microbes generally evolve with their hosts, but when microbes find their way to new environments, they often are more destructive.

miles within a day. The barriers that nature has placed can now be overcome in a day.

THE HISTORY OF INVASIVE MICROBES

Invasive microbes are microbes that have not naturally evolved in the environment where they are living. In theory, this would make them more aggressive and competitive. Defining a microbe as invasive is a bit tricky, however, because many microbes, specifically pathogens, are seen naturally as invasive or destructive to the host. Disease is a natural part of the Earth's ecosystem. Without any attempts to control disease, human populations would be periodically reduced.

The important distinction when defining a microbe as invasive is whether it has been introduced far from its place of origin, generally with much different hosts and a lack of factors to keep the microbe in control. One factor that could keep a microbe's population in check is another type of microbe. Competing microbes are known as **antagonistic** and are important in maintaining population control over an invasive microbe's growth.

Microbes are just as likely to cross borders and cause problems in a new environment. This is most easily seen by the spread of disease from European explorers, such as Christopher Columbus, to native peoples in the Americas. By the end of the fifteenth century, measles, influenza, mumps, smallpox, tuberculosis, and other infections were common in Europe. Explorers from these crowded, European cities brought these infectious diseases to the New World, where the microbes that caused the diseases did not yet exist. Because people in the New World had not ever been exposed to these pathogens, they were particularly susceptible. The first epidemics brought by the Europeans were the most severe. By 1519, smallpox appeared on the island of Santo Domingo, where it killed one-third to one-half of the local population and spread to other areas of the Caribbean and the Americas.[13] The population of central Mexico is estimated to have dropped by one-third in the single decade following contact with the Europeans in 1519.[14]

When humans travel, they can transport a pathogen on or in their body. Each individual has their own genetic makeup and immune response. Most times, the person has immunity to the pathogen they are introducing. In addition to the pathogen, the traveler may also be introducing the disease vectors, such as lice. A **disease vector** is an organism that by itself may not cause disease, but the microbes on it do cause disease. The microbe is unwittingly spread because the immune person is unaware that they are spreading the microbe.

SEA BALLAST WATER

Ballast water is a significant source of exotic species introduction. When a ship is empty of cargo, it sits high in the water and can be more easily overturned by wind or waves. To make travel safer, the ship's tank will be filled with millions of gallons of water—the equivalent of a small lake—to make it heavier and therefore more stable in transit along coasts and on the open seas. This water is referred to as **ballast water**. Water is collected from one port and then dumped in another.

The problem with ballast water is that species that are native to one port are suddenly the outsiders in another port thousands of miles away. Scientists estimate some 3,000 alien species per day are transported in ships around the world (Figure 2.2). These organisms range from microscopic plants and animals to crabs, mussels, and even schools of fish. Not all of these species survive, but the ones that do will produce offspring; the population then continues to increase exponentially.

The origin of many pathogens is traced to ballast water. *Vibrio cholerae* is a bacterium that causes a severe diarrheal disease

Figure 2.2 Ballast water contained in cargo ships is a significant means of transporting invasive species to new environments. Ballast water is used in a ship to weigh it down, but often this water contains organisms that will thrive in other regions where they have no predators.

called Asiatic cholera or epidemic cholera. This bacterium is one of the most common organisms in surface waters. It was introduced to Asia through shipping. Scientists have isolated this organism in ballast water of cargo ships from the Gulf of Mexico. Although cholera has been referred to in literature dating back to the Greek physician Hippocrates, this disease has been associated with numerous epidemics as a result of being introduced in a new location.

Vibrio cholerae is destructive to the human body because it produces cholera toxin. This toxin activates an enzyme in the intestinal cells that converts these cells into pumps that extract water and minerals, such as potassium, from blood and tissues. These fluids are pumped into the intestines, which results in massive diarrhea. A patient can lose gallons of fluid within a day or two and can die if no treatment is provided.

A MICROBE'S OPPORTUNITY

The world has changed much since microbes first existed. Human activities are the most potent factors driving disease emergence.[15] Not all microbes cause disease, however.

The environment is changing in ways that help microbes get around the globe. Pathogens have new opportunities to infect new hosts. Similar to all invasive species, invasive microbes move around because of travel. People living more closely together enhance the ability of microbes to spread. Breakdowns in public health measures, such as proper sanitation of sewage, also contribute significantly to the spread of microbes.

A microbe in a new region faces the same obstacles as an introduced animal or plant. The microbe must be able to survive in its new environment and reproduce; often this requires the microbe to enter a new host.

The type of host that a microbe uses reflects its ability to survive. Microbes that have specific environmental requirements, arthropod vectors (such as ticks), or complicated life cycles usually have a harder time establishing in a new environment. A complicated life cycle for a microbe would be one that requires many hosts from birth to reproduction.

Schistosomiasis (SHIS-toe-SO-my-uh-sis) is an example of a disease that would be difficult to spread to another country because of host requirement. Schistosomiasis, common in southern Africa, can be contracted when swimming or wading in waters that contain a parasitic worm. The symptoms of this disease begin with a rash followed by fever, chills, and muscle aches. The spread of this disease is somewhat controlled by the needs of the worm that causes it. This parasitic worm must have a suitable snail intermediate host that also exists in that region.[16]

In some cases, sanitation standards prevent the spread of microbes that would otherwise cause disease. *Taenia solium* is a parasite that is transmitted from human to human through eggs in feces. In the United States, toilets and other sanitary measures

Oil Spills and Wastewater: Beneficial Pollutant-Eating Bacteria

Bacteria can be used to clean up oil spills and wastewater. Bacteria feed on carbons, which are present in large quantities in wastewater and oil. These bacteria have been able to control carbon output by factories.

In Gary, Indiana, the U.S. Steel factory discharges large amounts of carbon daily into a nearby river. U.S. Steel exceeded the levels of carbon pollutants by six times the allowable levels during the summer months. Scientists worked with the steel company to add the right kind and quantity of bacteria to eat up the carbon. The fish needed to breathe oxygen through their gills to survive. Before the bacteria was added, the carbon compounds were consuming oxygen in the water, threatening the fish. Officials at U.S. Steel knew they needed to do something. The added bacteria were able to consume the oil and therefore make oxygen more available to the aquatic life in the river.

halt the spread of this parasite that in other circumstances might spread disease.

Landscape fragmentation is another cause for the increase of pathogens today. Landscape fragmentation occurs when land of one type, such as forest, is broken up to accommodate a variety of uses, such as farmland or development. This process brings animals, people, and vegetation all closer together, able to exchange microbes.

Microbes also now have the potential to kill more people than they did a hundred years ago. Part of this is due to an increase in population and the globalization that leads to travel and closer contact with wild animals as mentioned. In addition, because advances in science enable people to live longer and those with chronic illnesses can still thrive, many people are considered to be immunosuppressed. People with a suppressed immune system, whether it is due to age or illness, are more vulnerable to pathogens.

Exotic Pets

The international pet trade is a multibillion-dollar business. Exotic pets are often smuggled into the United States. One such pet is the Amazon parrot from Latin America. The smuggled animal may have a pathogen on or in its body. If the newly introduced pathogen is rare or unknown in its new environment, an epidemic could occur. Amazon parrots from Latin America often carry exotic Newcastle disease (END), but do not show symptoms. If these parrots have the virus that causes END when they are smuggled into the United States, they can shed it for more than 400 days.

Agriculture

In a small town in the foothills of Jamaica's Blue Mountains, every Wednesday is market day. The town is Ewarton, and people come from adjacent towns by foot, bus, and taxi to sell livestock, grain, toiletries, jewelry, and anything else you can imagine. A bucket of bloody chicken feet sits next to fish and dried beans. The tropical sun beats down on dogs and people passing in the street as vendors yell to them hoping to entice them to buy. Even though the market is outside, the smell is strong of fish and poultry, presented without preservatives except for salt and ice. These agricultural practices common to developing countries also bring potential hosts and pathogens closer together (Figure 2.3).

Agriculture in developing countries has also changed in a way that more easily develops new microbes and spreads old ones. The mingling of domestic animals and wildlife can spread

Figure 2.3 In many developing countries, open markets are common with people, plants, microbes and animals all coming together. This mingling encourages host and pathogen transmission. In nature, it would be unlikely for wildlife, such as a monkey, to be in close proximity to a domestic animal, such as a dog.

diseases. Normally, wildlife and domestic animals are not close enough to one another to trade pathogens, but in a developing country, this often occurs.

Scientists believe SARS originated in bats. It is possible that bats were kept in the same live-animal market in China where palm civets (*Paradoxurus hermaphroditus*), squirrel-like animals common in southern China, were also found to be carrying the SARS virus. A densely packed animal market, where wildlife, domestic animals, and humans mingle in less-than-sanitary conditions, provides the ideal situation for a virus to jump from one species to another.[17]

3 Whirling Disease

THE CASE OF THE TAIL-CHASING TROUT

John has loved fishing ever since he was a young boy. John's father first began teaching him to fish for carp in a stream by a wastewater treatment plant. The carp with the big ugly lips were always thrown back into the water and never eaten. That way, John and his father did not have to worry about what contaminants might be lurking in the wastewater treatment plant's effluent. Gradually, John and his father began fishing for trout in Colorado's Red Feather Lakes near Wyoming. Surrounded by mountains and sunshine, it was some of their best fishing. On days John could not make it to Red Feathers, there was always the fish hatchery down the road. What it lacked in pristine scenery it made up for in a reliable supply of trout in relatively clean waters.

John was completely shocked when he was fishing one morning and saw a deformed trout. The spine was twisted, making it difficult for the fish to swim forward. He watched more closely and saw another trout so deformed that its tail was practically touching its head, causing the fish to swim in circles. Polluted waters were John's first thought. He was wrong. Little did he realize that a parasite could cause such deformities.

Whirling disease is caused by a nonnative microscopic parasite that eventually destroys the nervous system of certain fish, specifically trout and salmon. Fish become deformed when infected by this parasite and often develop a black tail.

They exhibit the erratic tail-chasing behavior that gives the disease its name.

The microscopic parasite that causes whirling disease is actually an amoeba (*Myxobolus cerebralis*). The amoeba produces spores that make this disease profoundly difficult to control. The abundance of these spores makes eradicating the source of whirling disease nearly impossible in infected waters. The spores are also extremely small: more than 4 million spores can fit on the head of a pin.

THE PARASITE'S LIFE CYCLE

Like many parasites, the whirling disease parasite must live in two hosts to complete its life cycle. This bottom-dwelling parasite lives first in a worm that lives at the bottom of the water. Inside the worm, the spore changes form or metamorphoses and becomes a highly effective form, the **Triactinomyxon** or TAM. The TAM is then released from the worm and into the water where it is free floating. The TAM clings to the body of the fish and works its way into the fish's nervous system, where it reproduces rapidly. While inside the nervous system, the TAM metamorphoses again, this time into a spore. The spore moves to the fish's cartilage near the head.

After several weeks of this spore infestation, trout and salmon start to display symptoms. The spine is often deformed causing the whirling behavior. The infestation of TAMs also disrupts the fish's equilibrium, which adds to the erratic swimming behavior. The head may also be deformed from a shortened, twisted jaw. This is not the end of the cycle though. When the fish dies, the body breaks open and releases the spores, and the cycle starts all over again.

Whirling disease is devastating coldwater fisheries in North America. Trout and salmon have been the most notably affected, but other fish in the same family appear to be suffering as well, such as the mountain whitefish (*Prosopium williamsoni*). Rainbow trout (*Oncorhynchus mykiss*) and cutthroat trout (*Oncorhynchus clarki*) are also particularly vulnerable (Figure 3.1). Young fish tend to be more susceptible in general.

COEVOLUTION: A HOST'S DEFENSE

Whirling disease originated in Europe. Fish native to Europe, such as the brown trout, evolved along with it. When two species of organisms coexist and evolve in response to a similar environment, this is known as **coevolution**. As the parasite becomes more fit as a species, the brown trout through natural selection would also evolve to be more resilient to the parasite.

The classic example of coevolution involves plants and insects. A plant produces chemicals in its structure that are

Figure 3.1 Rainbow trout are a favorite fish among anglers in the United States. Whirling disease has had a devastating impact on the rainbow trout, particularly the younger fish. The parasite that causes this disease destroys the trout's nervous system, causing an erratic tail-chasing behavior.

poisonous to insects or, at the very least, unappealing. The insect bites a leaf and dies, or it is discouraged and searches for another plant. This is the insect's immediate response, the one that occurs on the individual level. Nature is not so easily deterred. In response to the plant's toxic compounds, the insects develop their own ways to detoxify these compounds, essentially developing a way to tolerate them. The plant responds in its own evolutionary way by developing new chemicals. Insects respond and on and on it goes. Taking an insect, plant, animal, or microbe from its natural environment to a new one, where members have never had to interact with it, is a potentially disastrous situation. The native animals would have missed out on decades of evolving to compete with the new organism. The invader has an unfair advantage.

The importance of coevolution is well illustrated by whirling disease in the United States. Trout and salmon native to the United States did not evolve with the parasite that causes whirling disease. The brown trout, native to Europe, did evolve with the parasite that causes whirling disease. So, even though trout species in general tend to be susceptible to this parasite, the brown trout has immunity as a result of coevolving with the parasite that causes whirling disease. Even though the brown trout does not develop the deformities and subsequent erratic swimming pattern from this parasite, it can still be a host. When the brown trout dies, thousands to millions of the parasite spores are released into the water.

CONTROL OF WHIRLING DISEASE

Whirling disease was introduced in the 1950s and it is now found in at least 22 states as well as several European countries. The best way to control whirling disease is to prevent its spread into new bodies of water. Since this parasite can occupy both the fish and water, depending on the stage of the life cycle, no part of a fish, sediment, or water should be transported. Transporting fish from one body of water to another is illegal in most states.

Stocking fish spreads this disease. In Colorado, the introduction of whirling disease is traced to the 1980s from imported trout from a private hatchery. The natural migration of fish also contributes significantly to the spread of whirling disease. The main reason that stocking and migration cause new incidences of whirling disease is because the fish bodies usually contain the parasite's spores. This is true even if the fish does not display whirling disease symptoms.

Spores are contained in the body and water of infected areas. Spores are often the pathogenic aspect of many microbes. These spores are nearly indestructible, as they can survive freezing and drying. In a stream, they can survive anywhere from 20 to 30 years. The longevity of the spores is why it is so crucial to prevent infected waters from contaminating clean waters.

The disease is also spread by birds that drink water from infected areas and then excrete the viable spores in another body of water. Anglers and boaters may transport live fish from one body of water to another, which spreads the disease. Anglers may use infected fish as cut bait and then unknowingly spread the disease (Figure 3.2).

Anglers are warned not to dispose of fish parts in the water after they have cleaned their fish. Boaters are told to rinse all mud and debris from their equipment and gear before going to new waters because spores could be contained in the debris. Boaters must also drain water from their boats before leaving.

Hatcheries should be tested for whirling disease. Testing is very accurate; waters that contain as few as two spores will test positive. In Colorado, eight of the state's fish hatcheries tested positive, and measures were taken to reduce the disease in hatcheries. Infection appears to have been reduced in hatcheries that tested positive, and monitoring will continue.

Figure 3.2 Spores are the source of infection for whirling disease. Even if diseased fish are removed from waters, the spores can survive for decades and spread whirling disease. Anglers have unknowingly helped to spread the disease by using infected fish as cut bait.

Killing TAM (Triactinomyxon)

Anglers fishing in heavily TAM-infested waters are encouraged to take extra steps to prevent the spread of whirling disease. Rinsing and drying boots, waders, and any fishing equipment will generally kill any stage of TAM.

Chlorine, such as household bleach, can also be used to kill TAM at any stage. Mixing 1 part chlorine with 32 parts water and keeping it in contact with infected materials for 10 minutes will ensure that the TAM is killed. The mature spore is more difficult to kill. The mature spore, or **myxospore**, is found in the mud of an infected body of water. Mixing 1 part chlorine with 9 parts water and soaking for 10 minutes will ensure that the mature spore has been destroyed.

State and local officials can provide clean water and a hose at boat ramps to encourage people to wash off before leaving a stream or lake. Educating people by providing information about whirling disease and posting maps of known locations of infected waters can help raise awareness while encouraging cooperation and compliance to stop the spread. People also need to know what they can do to cleanse themselves and their equipment of the parasite. Posting information at recreational areas as well as including information in angler and boating publications and visitor's centers are great first steps.

Fisheries experts in Colorado are using ultraviolet light to kill spores in hatcheries. While the success of these measures is not yet known, this kind of control without using pesticides is appealing to environmentalists.

States with infected waters are coming together to promote research about whirling disease. National conferences help natural resource managers share methods of control while ensuring that research is not being duplicated.

Sudden Oak Death

FEAR THE FUNGUS

4

Phytophthora ramorum, the fungus responsible for oak death in North American forests, causes disease on the aboveground portion of trees. This fungus does not just live in oaks as the name implies, but also on other tree species. *P. ramorum* can infect dominant mature trees, shorter, shade-tolerant trees, or trees in streamside habitats. Because this parasite is a fungus, it produces spores. Both evergreen and deciduous trees can be hosts.

PLANTS AND DISEASE

Grasses, broad-leaved plants, shrubs, and trees are all plants. Plants are considered healthy when they are able to complete the normal processes that enable them to grow and reproduce. These processes include the absorption of water and nutrients from the soil, production of seeds and spores, and photosynthesis. **Photosynthesis** is how a plant feeds itself by using the energy in sunlight to make sugar out of carbon dioxide and water. A plant disease is the response of plant cells and tissues to a pathogen or environmental factor that results in adverse changes in the form, function, or integrity of the plant. The disease may lead to partial impairment or death of the plant.[18] Symptoms of plant disease are wilting, rotting, and discoloration.

SYMPTOMS OF SUDDEN OAK DEATH

The symptoms of sudden oak death are subtle. The leaf tips and edges turn yellow. A yellowing of the leaves is known as **chlorosis**. Infected leaves of evergreen shrubs readily fall off, which creates a forest floor thick with needles. Symptoms of sudden oak death appear as large cankers (Figure 4.1). Cankers are places on the tree, such as the bark, where the tissue has died and eventually rots. **Cankers** usually appear sunken because the surrounding tissue continues to grow. Dead shoots may also appear from the canker.

The tricky thing about sudden oak death is that it is difficult to isolate the pathogen. Isolating the pathogen by taking a plant sample and analyzing it in a lab is the only way to know for sure that a plant is infected with sudden oak death. Because the symptoms described above are not unique, scientists need to do the testing to determine how to manage the disease. It is common to have a false-negative result when testing for this fungus. Scientists have learned that prolonged soaking in water is necessary to isolate the pathogen from some plants.

The Fungus and the Forest

Controlling an invasive fungus begins with knowing where it is. Detection is a significant challenge with invasive microbes. Looking for the invasive brown treesnake is much easier than detecting an invasive organism that cannot be seen with the naked eye. According to the U.S. Forest Service, it is like looking for "a needle in a haystack."[19] If scientists suspect a tree is infected with *P. ramorum*, a sample is taken and watched in a lab culture to see if the fungus develops (Figure 4.2).

One of the main challenges to mapping sudden oak death is that the symptoms described above are not unique. The leaf spots, discoloration, and cankers associated with *P. ramorum* differ only subtly from symptoms caused by other leaf-inhabiting

Figure 4.1 Cankers are a sign of sudden oak death. A canker occurs when dead tissue sinks away and rots. Cankers appear sunken because the surrounding tissue continues to grow.

or canker-causing pathogens.[20] In forests that are infected with the fungus, patches of dead or dying trees usually appear on a rugged, heavily vegetated terrain that can be difficult to get to.

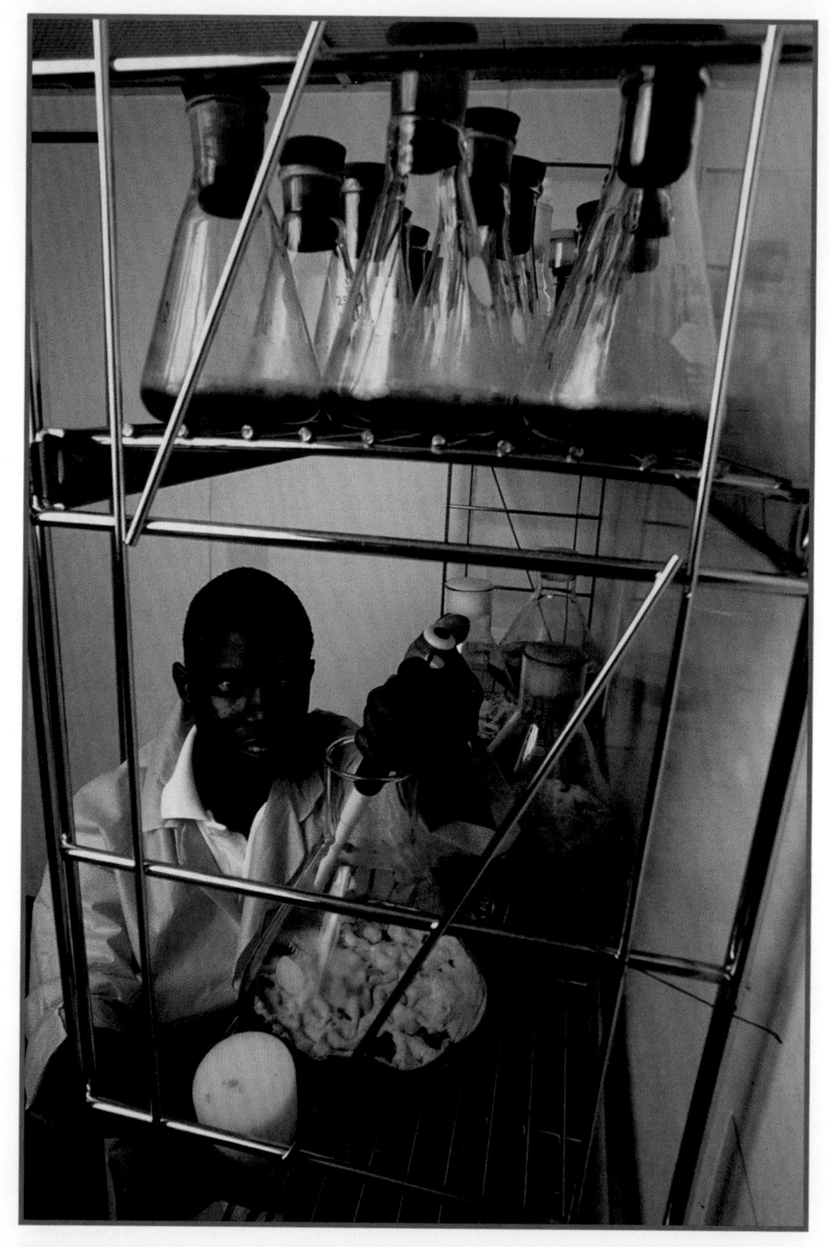

Figure 4.2 Scientists take a sample from a tree if it is suspected of being infected by the fungus that causes sudden oak death. The sample is then grown in a laboratory to confirm the presence and identity of the fungus.

Even with all of these difficulties in mapping the fungus, natural resource managers are surveying millions of acres of forests in California and Oregon. Along with determining the extent of the problem, surveyors will note the impact as well. A few different ways to map the disease will be used to come up with a best guess of areas affected with the fungus (Figure 4.3).

Small fixed-wing aircraft are used to do aerial surveys. The advantage of using aircraft is that all types of terrain can be accessed, unlike ground surveying where there are limits to access. Two observers fly in the aircraft at about 1,000 to 2,000 feet (300 to 600 meters) above the ground and map out areas where they see dead trees.

The best way to get a high degree of accuracy when trying to determine an exact loccation on Earth is through global positioning systems (GPS) and geographic information systems (GIS) technologies. GPS refers to the hardware that is used when a person is outside collecting data. GPS equipment usually consists of a handheld device that collects electronic data by pinpointing the location using satellites. GPS data downloads to a map that shows the geographically referenced locations. A program that is able to read the data that comes from a GPS unit is referred to as GIS. The program can interpret geographically referenced data, meaning that each data point collected in the field would be associated with geographic information, such as latitude and longitude.

When managing invasive species, it is important to have detailed information, such as latitude and longitude, so that people can return to the recorded location to determine if a management plan is working. The GPS data collected in the field is usually integrated with a map to show the distribution of infected trees.

The observers in the aircraft rely on typical crown shapes and discoloration from sudden oak death. The foliage will

Figure 4.3 Learning to use a map and compass are great tools for becoming a natural resources professional. Mapping is used to manage sudden oak death by first determining how many acres have been affected. Mapping also helps determine the range and degree of sudden oak death.

appear reddish-orange to brown or will simply be dead. Despite all the difficulties in the unique appearance of symptoms, aerial surveys have been successful. Tan oaks (*Lithocarpus densiflora*), a Pacific Coast evergreen tree, also die from *P. ramorum.* In Oregon during a special aerial survey, some dead and dying tan oaks were spotted in July 2001. Approximately 330,000 acres were surveyed that same month and then another half a million acres in October. The two surveys revealed 41 occurrences of dead tan oaks. Samples of the trees were taken and cultures were grown in labs, but only nine of the sites proved positive for *P. ramorum.*[21] All infected trees were entered in a database with their associated geographic location for monitoring.

A more tedious method is the ground-based survey design; staff walks the grounds marking on a map where they believe infected areas are. Ground surveys are often used to back up aerial surveys. In places where observers in aircraft claim to see dead or dying oaks, ground crews will verify this information. Also, understory trees can best be evaluated with ground surveys because observers would not be able to see the shorter trees from the air. Ground surveys also provide an opportunity to take samples to be cultured in a lab.

Scientists are also looking into using models and image analysis. Image analysis consists of taking a picture from above that can determine which trees are infected. Computer models plug in different variables to enable the computer to determine exactly where the fungus is located in the forest. One variable that could be used in the model would be a likely pathway for movement of the fungus and another variable could be favorable climatic conditions.

SUDDEN OAK MANAGEMENT

Integrated pest management (IPM) is the best strategy for controlling *P. ramorum.* IPM controls plants, animals, and microorganisms by using chemical, mechanical, cultural, and biological methods. Chemical control methods are pesticides; mechanical controls include erecting barriers to invasive fish in streams or mowing weeds; cultural methods are strategies that are used on a repeated basis, such as flooding to kill plants or animals; biological control introduces a natural predator to control the pest. By using more than one control strategy, the expectation is that the chance of success will be greatly improved. As with any invasive species management program, however, each situation has different variables. In some cases, the best strategy will involve using a few different tools. In others, one specific method may be more effective than others. At a minimum, IPM requires land managers to examine all of their options.

Recommended IPM strategies for sudden oak death include cultural control and host resistance. Control strategies can, however, be any of the above-mentioned methods and are generally repeated practices to control the unwanted species. Control strategies are well-suited to a nursery because they can be implemented along with automatic tasks, such as cleaning the windows and taking plant inventory.

Water management is a critical component in controlling *P. ramorum* because water is a source of spread for sudden oak death. Spores need a wet period to form sporangia, which is the capsule that contains the spores (Figure 4.4). These sporangia are the source of new infections. *Phytophthora ramorum* spores need to stay wet for 8 to 12 hours. Irrigating in the late afternoon will allow enough time for susceptible plants to dry off.

One cultural control strategy in a nursery involves designing the layout of *Phytophthora*-susceptible crops in a way that avoids standing water and the potential for splash dispersal during irrigation.[22] Standing water can be avoided by crowning nursery beds in the middle so that the water rolls off the edge of the beds. Gravel and porous fabric used as ground cover also help to avoid water puddles around pots. Ditches are often used in nurseries to return irrigation water. When the ditches are planted with vegetation, this will help slow the return of water and therefore the amount of stagnant water.

Knowing the species, susceptibility, and age of plants in the nursery will help minimize loss from this fungus. All fungus-susceptible plants should not be put in the same location in case an outbreak does occur. Instead, plants that are not susceptible to the fungus, or nonhosts as they are known, should be interspersed with susceptible plants to slow the spread.

Figure 4.4 Rust spores are visible on the leaves of this yellow starthistle plant. Spores are the reproductive agents for protozoa, fungi, and ferns.

Another major issue of water management in nurseries is the recycling of water with the flow in and out of a retention basin. Irrigation water that becomes contaminated from infected plants is pumped back to a basin of water or retention basin through a pipe. Water that flows through a pipe instead of percolating through the ground does not have a chance to filter out pathogens. Therefore, recycling irrigation water in a nursery continues the dispersal of fungal pathogens. Water can be treated through chlorination, ultraviolet (UV) exposure, and slow sand filtration to cleanse the water of the pathogen. In UV exposure, water passes over a fluorescent light. The light penetrates the microbes enough to mutate the part of the organism that enables it to grow and reproduce. Ultraviolet exposure may not always kill an organism right away, but it will prevent its spread because it will not be able to reproduce.

Host resistance is another tactic used in nurseries to control the loss of plants from *P. ramorum*. Host resistance involves a plant that does not succumb to disease from the fungus. A naturally resistant plant would be a good choice for a nursery that has a problem with *P. ramorum*. A cultivated variety of rhododendrons named Cunningham's white is known to be resistant to this pathogen. Horticulturists can screen each variety of nursery plants to determine those that are naturally resistant.

Biological Control

Biological control is a method of introducing an invasive species' natural predator to control it. The concern is the potential of the introduced species to become a pest. Early in the days of biological control, such mistakes were made. Now, introduced species undergo lengthy experiments by federal agencies to make sure they will not adversely affect native species.

The cane toad (*Bufo marinus*) was introduced worldwide to control pests of sugarcane. True, the cane toad did eat pests of sugarcane, but it also ended up eating just about anything else in its path. Cane toads have been reported eating rotting garbage, fledgling birds, and even a lit cigarette butt.

If these toads had remained focused on consuming the pests that they were intended to feed on, it would have been an immensely successful biological control campaign. Because these toads are not **host-specific**, however, there has been a decline in native species, such as lizards and other amphibians. A host-specific organism will typically have very tailored feeding habitats, generally consuming only one species. In addition to the loss of organisms that it ate, this toad displaced native amphibians by competing for food and breeding habitat.

Even microbes have parasites that limit the microbial population. Using these other organisms to control an unwanted microbe is biological control. Biological control of *Phytophthora* species consists of manipulating microbial antagonists at the site of the infection. An antagonistic relationship is one in which one organism works against or compromises the livelihood of another organism.

Antagonistic bacteria and fungi produce antibiotics or toxins that inhibit growth and spore germination, or compete with the pathogen for a needed resource. Using these antagonistic microbes can be a biological control method of a pathogen such as *P. ramorum*.

Introducing the antagonistic microbes is not enough. A specific level, or threshold, must be maintained to have success controlling *P. ramorum*. The **threshold level** is a specified population size of the microbes to control the pathogens. If the number of antagonists falls below this threshold level, the pathogens will resume enough of their behavior to make disease likely to occur in the host. When the threshold level is maintained or increased, the antagonist's effects are displayed in the reduction of sudden oak death in susceptible trees. Populations of antagonists are supported by soils with high organic matter because they serve to nourish these microbes.

When people get sick from mold in the home, the spores are to blame. If the mold did not produce spores, the mold itself would pose little threat of an airborne illness. *Phytophthora* species also are suppressed when coming into contact with the spores of the pathogen.

Induced resistance is another biological control option for the pathogen that causes sudden oak death. Instead of weakening the pathogen, control can be achieved by strengthening the host's ability to fight the pathogen. This is the idea behind induced resistance. This biological control strategy injects microbes that stimulate a host's defense reactions into the belowground or aboveground part of the plant. The plant responds by alerting its defenses, making it more protected against pathogens.

Nursery Detection

One important place to detect the fungus is in nurseries (Figure 4.5). Reliable detection depends not only on proper knowledge about the pathogen itself, but also nursery practices. The best approach is to sample where the fungus is most likely to be growing in the field. Like most fungi, *Phytophthora* species need water to develop and infect. With this in mind, moist areas are the best places to look in the field. Plants in wet areas will show disease symptoms first.

A sample of the plant is analyzed in a lab to determine whether it is infected. The fungus is more easily detected in the twigs than the leaves, so that is where the sample is taken. It is best, however, to take the entire plant to the lab for sampling.

Figure 4.5 A plant nursery is a common place for fungi to spread among plants. Sanitary nursery practices and early detection can help prevent the spread of pathogens. The challenge to identifying a specific disease is the fact that symptoms such as rot or discoloration are typical of many pathogens rather than just one.

The challenge in a nursery is that many plant species are growing side by side, and *P. ramorum* causes different symptoms in different plant species. Nurseries are also usually fighting a variety of plant diseases, so again it is difficult to diagnose *P. ramorum* infection. Brown to black discoloration of single nonwoody twigs is characteristic of not only *P. ramorum*, but also other *Phytophthora* species.

Ralstonia Solanacearum

A BACTERIAL INVADER SUCCESS STORY

5

Ralstonia solanacearum is a bacterial pathogen that threatens important U.S. crops, such as potatoes, tomatoes, peppers, and eggplant. This pathogen causes southern wilt, bacterial wilt, and brown rot of potatoes.

In December 2003, *R. solanacearum* was found in a U.S. greenhouse. Geraniums imported from a greenhouse in Kenya with unsanitary practices were the source of the pathogen, although this bacterium is believed to originate in the highlands of Peru (Figure 5.1). Geranium shipments from Kenya were immediately stopped, and all existing geraniums were destroyed. The immediate concern was how much contamination was in plant material.

SYMPTOMS OF *RALSTONIA* INFECTION

Bacterial cells often contain a "master switch," a protein that commands other proteins to attach to plant tissue and, in some cases, to release toxins. These toxins cause disease in the plant. Wilting is a common symptom of many plant diseases. In a *Ralstonia solanacearum* infection, wilting is particularly pronounced on the lower leaves. This pathogen disrupts the part of the plant that is responsible for bringing water and nutrients up through the roots and out to the leaves of the plant. The sign that the vascular system (conducting system) has been disrupted by this pathogen is the roots turning brown.

Figure 5.1 A shipment of geraniums from a nursery in Kenya is believed to have been the source of an *R. solanacearum* infection in 2003. *Ralstonia solanacearum* is a bacterial pathogen that infects important U.S. crops, such as potatoes and tomatoes.

To determine if the wilting is from *R. solanacearum*, scientists take a sample of the dying plant and analyze it. One way to analyze a plant sample is to take a cross section of the stem base, place it in a tube of clear water, and look for white milky strands. These strands are the bacterial cells of *R. solanacearum.*

Ralstonia and Bioterrorism

Ralstonia solanacearum can be spread through soil, irrigation water, and gardening equipment. Taking a cutting from an infected plant can contaminate a knife that may be used on other plants. This pathogen does not, however, readily spread from plant to plant through the air or casual contact.

Figure 5.2 Potato brown rot as depicted in this photo is caused by *Ralstonia solanacearum* bacteria. Bacterial cells of *R. solanacearum* instruct other proteins to attach to plant tissue and release toxins. These toxins cause the wilting seen here.

R. solanacearum is of particular concern because of the way it spreads and its ability to affect the U.S. food supply. This pathogen is listed in the Agriculture Bioterrorism Act of 2002. Government officials fear that *R. solanacearum* could be intentionally used to destroy the U.S. potato industry. If a potato

Agricultural Bioterrorism: Invasive Species As a Weapon

Bioterrorism is the act of intentionally creating and dispersing microbes to inflict harm or death to people in an act of aggression. Terrorists might introduce an invasive microbe, specifically a pathogen such as the Ebola virus or the smallpox virus, to weaken the will of the American people. Invasive species, whether they cause disease or not, are a huge financial burden that harms the U.S. economy. An outbreak of disease from an introduced pathogen would also divert military personnel in the case of an extreme emergency. Military personnel would be needed to control the invasive species by enforcing a quarantine or caring for the sick.

Car bombs and subway attacks are meant to instill a low level of constant fear in the minds of the terrorist's targets. Experts believe that the long-term economic, health, and psychological effects of using invasive species could strike a tremendous blow to the United States by exhausting resources and national will over time.*

Experts on bioterrorism are concerned that research used to find cures for diseases may be used to pose a threat to the public. One tactic that a terrorist could use would be to acquire strains of a particular pathogen that are resistant to the available antibiotics. Most of the infections acquired in hospitals, which are easy places to get an infection, are resistant to at least one antibiotic. The U.S.

became infected with this bacterium, it would develop brown rot (Figure 5.2), which could inadvertently be spread.

Some scientists in Wisconsin argue that *R. solanacearum* would not be a likely bioterrorist tool because it can only spread through soil or water, not air. These scientists want *R.*

government and concerned scientists are looking into ways to protect this information so that it will not find its way into the hands of terrorists. One suggestion is to have a certification requirement for lab technicians working on pathogens.

Terrorists could also release an invasive microbe to compromise the food supply in the United States. The nation's food supply is vulnerable to an attack because it is grown in concentrated locations. California produces 100% of the almonds grown in the United States. A terrorist would only have to distribute a pathogen among almonds grown in California to nearly destroy the nation's supply of almonds.

How easy would it be to accomplish bioterrorism with an invasive microbe? Unfortunately, it would be fairly simple because invasive species are cheap and easy to produce, acquire, and introduce. Populations of microbes grow exponentially, so a terrorist could introduce the invasive microbe in a few different locations to maximize damage. Biologists with as little as $10,000 worth of equipment could produce a significant quantity of biological agents.

***Robert J. Pratt. "Invasive threats to American homeland." *Parameters.* (March 22, 2004). Available online at http://www.highbeam.com.**

solanacearum to be taken off the bioterrorism list; the federal restrictions on conducting research with pathogens on the list are so cumbersome that many researchers decide to avoid working with those listed.

Caitilyn Allen, a plant pathologist at the University of Wisconsin-Madison, studies *R. solanacearum* and had to get a permit to continue her research. Allen and her fellow researchers were also subjected to an FBI background check. In addition, Allen had to spend more than $50,000 to set up a second high-security lab apart from the general work area. All this high security does not make much sense to Allen,

Who Controls the Movement of Plants Through Borders?

The U.S. Department of Agriculture's (USDA) Animal and Plant Health Inspection Service (APHIS) is responsible for ensuring that plants that enter this country are free from pests and diseases. APHIS inspects plants that are mailed, carried, or shipped into this country by travelers and nursery owners.

APHIS inspectors question every international traveler going through customs about whether they are carrying plant parts in their baggage. It only takes the actions of one person to introduce an invasive species that could have devastating effects on our environment.

APHIS staff looks for signs of invasive microbes, usually in the form of a plant or animal disease, at ports of entry. At these plant inspection stations, inspectors work with specialists in the fields of entomology, botany, and plant pathology to locate, examine, and identify exotic pests, diseases, and noxious weeds. Entomology is the study of insects; botany is the study of plants; and plant pathology is the study of plant diseases.

given the fact that *R. solanacearum* is still widely available in Europe and in tropical countries.[23]

Prevention and Control of *Ralstonia*

Since the 2003 incident of contaminated geraniums finding their way to a U.S. greenhouse, the U.S. government has required that all geranium imports from countries with *R. solanacearum* be tested and certified free of this bacterium. They must meet production facility sanitation requirements.

Quarantine is the primary method for controlling the spread of *R. solanacearum*. A quarantine consists of inspectors placing

APHIS is not trying to eliminate the introduction of foreign products that could be vectors (carriers) for invasive microbes. Many of the fruits and vegetables we eat every day have been imported from other countries. APHIS ensures the safety of these imports with a few precautionary steps.

Importers must apply for an agricultural import permit and secure a phytosanitary certificate from the exporting country. Phytosanitary certificates verify that plant quarantine officials from the exporting country have examined the plants for pests and diseases prior to exporting them. Once the plant arrives at a U.S. port, an inspector examines samples from each species of plant and seed. The inspection process includes a meticulous examination of the leaves, stems, roots, and seeds of the plant. If an inspector discovers a disease, they determine the extent of potential harm and make a decision to quarantine, export, or destroy the plant.

orders on the plants they suspect are infected to prevent any sale or distribution of the potentially infectious plants. The plants are then stored until testing of the plant determines if it contains the *R. solanacearum* bacterium. If plants test positive for the bacterium, they are destroyed. Any growing facility that came into contact with the infected plants is ordered to disinfect.

The outbreak that resulted from the Kenyan geranium imports occurred in 4 Midwest greenhouses. Plants were held in more than 900 greenhouses in 47 states. Plants in 127 greenhouses in 21 states tested positive for the disease; some 2 million geraniums were destroyed.[24]

Exotic Newcastle Disease

6

GASPING CHICKENS

In San Diego, California, eggs are big business. Of all the money generated from egg and poultry products, 94%, or $48.7 million, comes from eggs. This could all change because of a recent outbreak of exotic Newcastle disease (END).

Exotic Newcastle disease is a highly contagious and fatal disease caused by a virus found in birds. Of all bird illnesses, END is one of the most infectious. Exotic Newcastle disease affects the respiratory, nervous, and digestive systems of all birds, although the virus is so strong that many birds die before showing any signs. The death rate for birds that contract the virus is nearly 100%. Currently no cure or approved vaccine exists.

In San Diego County, END was discovered in December 2003; it has since spread from private farms to six commercial ranches and ten small poultry operations. When a California State veterinarian saw two dead chickens brought in to his lab from separate backyards, he knew something was terribly wrong. These backyard farmers were seeing their flocks die quickly and wanted to get a diagnosis.

Farmers in San Diego whose poultry stocks test positive for the virus must kill their entire flock. Although the federal government reimburses farmers for birds they must euthanize, the situation is still financially damaging. Once all of the birds are killed, the farmer must disinfect the farm, obtain certification or approval from the USDA that the farm is free of the virus, buy

Figure 6.1 In the United States, one of the ways to control exotic Newcastle disease is to kill infected chickens. Chickens with exotic Newcastle disease will transmit the disease to other chickens through bodily discharges.

chicks all over again, wait for them to mature, and then finally cash in on the eggs they produce. This means that the farmer has to wait six months or a year before any money comes in.

SYMPTOMS AND SPREAD OF EXOTIC NEWCASTLE DISEASE

The incubation period for this virus is 2 to 15 days. An infected bird is likely to gasp for air, cough, have drooping wings, twisting of the head and neck, and greenish, watery diarrhea. Its head may actually swell.

END spreads mostly through the bodily discharges of infected birds (Figure 6.1). Bodily discharges include bird droppings and secretions of the nose, mouth, and eyes. The virus is alive and active in these discharges and can then be picked up by people's shoes to infect a flock in another location. The END virus can survive weeks in a warm and humid environment on birds' feathers, manure, and other materials. The virus can survive indefinitely in frozen material. Fortunately, dehydration and ultraviolet radiation destroy END.

Debeaking is a common practice on poultry farms where the beak is removed from the bird so that birds in tight quarters cannot injure each other. In the process of debeaking, bodily fluids, such as blood, can be transmitted to other birds from the equipment used to remove the beak. This common practice contributes to the spread of END.

Prevention of Exotic Newcastle Disease

The USDA has a three-tiered approach to controlling END: destruction of infected flocks, quarantine, and surveillance programs. In California, 3.5 million commercial and backyard poultry, such as geese, chickens, turkey, pigeons, and peacocks, were euthanized to stop the spread of a recent outbreak.

Once an outbreak has occurred, a quarantine on poultry is enforced. Quarantine means that poultry farmers are not allowed to move their birds from one area to another. This helps prevent the spread of the disease once birds are infected. Compliance of private owners of poultry farms or backyard operations is a bit more difficult because often they do not know a quarantine is in force (Figure 6.2). Even so, quarantine is an honor system, meaning that authorities rely on the honesty of farmers to comply. There are no chicken police out on the roads to make sure farmers do not transport birds to new locations.

Halting Exotic Newcastle Disease

Although scientists are working to develop END vaccines, there are currently no vaccines available. Vaccines now being studied work by injecting the bird with a weakened live virus. The immune system is stimulated to produce antibodies to fight the weakened virus. The intent is that the bird's body will produce enough antibodies to fight infection the next time it is invaded by the END virus. The risk is that, by introducing a weakened virus, it is still possible to cause a deadly infection. Rather than being protected from END, some birds have died after receiving the test vaccine.

Scientists at the USDA are developing a better approach. A new kind of vaccine, called a **virosome vaccine**, eliminates much of the potential danger of END vaccines. A virus is dangerous to a host because it invades a host's cells and directs them to replicate or make copies of the virus. A vaccine currently being developed works by turning off the replicating characteristic of the virus. This greatly reduces the chances that the bird will develop the disease and die as a result of the vaccination. Scientists genetically altered the virus to stop its replicating behavior once inside the host's cells. The resulting virus is genetically different and does not occur in nature. When an

Figure 6.2 Crowded chicken farms are an easy way to transmit pathogens. Once a farm has been infected, a quarantine is often necessary. During a quarantine, an individual who might have contact with poultry outside the infected farm would not be allowed to enter.

organism is altered genetically this is referred to as **biotechnology**. As in the case of the virosome vaccines, biotechnology can be extremely useful, but it also has many controversial aspects.

Another way to contain END is by allowing only essential workers on the premises of poultry farms. Preventing the number of people in and out of farms, helps prevent the disease from spreading from one location to another. The USDA and the California Department of Food and Agriculture are considering putting armed guards on quarantined poultry farms. Given the millions of dollars that even a small outbreak costs the poultry industry and federal government, it just might be worth it.

Saving the Soybeans from Rust

7

Todd woke up one morning to the sound of his neighbor at his door. It was early, even for farmers in the state of Bahai in northeastern Brazil. His neighbor wanted to show him his farm. In nearly 13 years of farming, Todd had never seen anything quite like it. His neighbor's soybean field was just green stalks. Just days ago, the soybean plants were full of leaves.

Todd, a North Dakota native, and his neighbor had noticed a sky full of red dust. They later found out that the dust was actually spores, which caused the destruction of his neighbor's soybean field. Todd knew it was only a matter of time before the rust he had only previously heard about would devastate his own crop. This rust was Asian soybean rust (Figure 7.1).

Asian soybean rust is caused by a fungus called *Phakopsora pachyrhizi* that draws nutrients away from the soybean plant. A farmer with soybean fields containing this rust will have a lower crop yield.

Todd quickly went out and bought some fungicides. He applied them twice in that season, but still had a crop yield that was 40% less than in rust-free years. Many farmers in northeastern Brazil suffered similar losses. Unfortunately, the Asian rust that Brazilian farmers are so familiar with has now made it to the United States.

Soybean rust was first detected in North America in November 2004. This disease was first observed in Japan in the

Figure 7.1 Spore-producing fungi cause Asian soybean rust. Once infected, the leaves of a soybean plant will appear to have yellow spots that turn reddish-brown.

early 1900s and has since spread throughout Asia and Australia. It reached Hawaii in the 1990s.

A MULTITUDE OF HOSTS

Nearly every plant has a rust species that can infect it. The fungus that causes the rust has a wide host range. Remember that a host is an organism that helps a microbe complete its life cycle. A host range is all of the different kinds of plants that the specific rust can live on. For most rust fungi, the host range is usually a couple of species of plants. In the case of the Asian soybean rust, there are many hosts, 95 in fact, although many are not native to the United States. If any of these other hosts are harboring the same rust near soybeans, a new infestation can occur. Some common crops that can have soybean rust are clover, lima beans, and black-eyed peas.

Scientists believe the wide selection of hosts is a survival strategy for this fungus. Some diseases can overwinter, meaning that they do not die but go dormant. No symptoms would be displayed during the winter, but spring can renew the microbe's growth, causing disease. In the case of the Asian soybean rust, this fungus is not able to overwinter and therefore must go from one live host to another to survive and reproduce. Having more options of plants to infect increases the chance of having an available host, even in the winter. Fungi produce many spores; if one spore happens to fall on one of the many hosts available, the Asian soybean rust can begin to grow. Luckily, none of these other hosts develop a disease, as the soybean does.

As with most spores, those that carry Asian soybean rust are extremely hard to destroy and can withstand freezing. They travel long distances, as far as 500 miles (800 kilometers), by wind. Extreme weather events like a hurricane can make the spores travel more than 1,000 miles (1,600 kilometers). The combination of many hosts and hardy spores make Asian soybean rust a big problem for the soybean industry.

As with many microbial diseases, the symptoms display the mark of other pathogens. The rust begins as water-soaked regions and develops irregularly shaped spots that turn reddish-brown. Brazilian farmers have described the disease as appearing to spread from the bottom of the plant up. The underside of the leaves usually have more lesions or diseased areas.

Proper diagnosis is critical. Once the soybean plant is infected with the fungus to the extent that rust appears, the farmer has about 9 to 10 days to treat the plant before it completely dies. Detecting the rust this quickly can be quite an arduous task for farmers with thousands of acres. Farmers would have to walk their fields every day just to find the rust.

Prevention and Management Strategies

Willy is a Brazilian farmer who sprays his soybean fields with a fungicide to prevent Asian soybean rust. This fungus is so widespread in Brazil that Willy knows his crops would be damaged if he did not spray three times during the season. This is enough to prevent widespread crop losses.

Much of what we know about controlling soybean rust comes from Brazilian farmers who struggle with this fungus. Brazil's tropical climate is favorable to the production of the spores and ease of transfer among hosts since there is always something growing. American farmers are hoping early detection of the fungus will contain it enough to avoid having to add it to the list of crop pests to control annually.

Genetically Modified Organisms: A Solution to the Pest Problem or Frankenstein Plants?

Genetically Modified Organisms (GMOs) are organisms with intentionally modified genetic material. The process is called genetic engineering. Scientists alter an organism's genes to improve it or make it more useful to people. These improvements can include making the organism more competitive, virulent, or resistant to pesticides.

Genetic engineering is common in agriculture. Roundup-ready soybeans are one type of GMO. Roundup is a herbicide that is applied to agricultural fields to control weeds. Roundup works well to kill unwanted plants, but unfortunately can injure desirable plants as well. Once fields have been planted and the soybean plant has started to grow, application of Roundup would injure the soybean plant. Farmers were faced with a dilemma.

Scientists decided that if a soybean could be produced that would be resistant to the Roundup, then farmers could spray their

Because soybean rust is not widespread in the United States, the focus of controlling this disease is on early detection and treatment. The United States Department of Agriculture (USDA) has funded a project to help farmers by alerting them to new locations where soybean rust has been detected.

Spore traps can be used to detect Asian soybean rust spores. In July 2005, Asian soybean rust spores were found in a spore trap in Tennessee. No indication of the disease existed in the state, even though lab tests confirmed that the spores were in fact Asian soybean rust. The spores found in the traps were few, indicating that the spores had probably blown in from the Southwest. This information was helpful because it enabled the USDA to put farmers on alert for the disease.

weeds without worrying about injuring the soybeans. Scientists modified the genes of the soybean to produce Roundup-ready soybeans.

Genetically modified organisms are controversial. The benefits are obvious, but some environmentalists are concerned that these modified traits will enter nature's gene pool and possibly harm native species. If the gene that makes soybeans resistant to Roundup was to somehow be incorporated into the genetic makeup of an invasive plant species, that invasive would then be resistant to the weed killer Roundup. This situation will make it more difficult to control the invasive plant because the herbicide could not be used. Another concern is that a terrorist could use the altered gene to purposefully create a more competitive invasive species. Either scenario would have environmentally and economically damaging results.

The Internet also provides real-time information that allows farmers to get the latest alerts at the county level of confirmed reports of soybean rust locations. The Web has the latest advice from experts on managing the disease. Best farming practices are encouraged to target the fungus and minimize the amount of pesticides the farmer uses. Detecting the fungus before it has infected many plants enables the farmer to get rid of it entirely with a fungicide. If all small infestations are eradicated, then the farmer will not have to worry about spraying a fungicide every year as a preventative measure.

Offering information online is one way to get the word out to farmers; educational seminars are also helpful, particularly for low-tech farmers. Purdue University in Indiana and a local growers group sponsored free, educational meetings from January through March 2005, basically just before the growing season. Asian soybean rust is not yet present in Indiana, but with reported cases just two states away, farmers are arming themselves with knowledge to fight the fungus. The educational meetings help farmers to detect the fungus immediately and to properly treat it.

Bird Flu

8

In Jakarta, Indonesia, a woman is feverish and has difficulty breathing. In a congested city such as Jakarta, it is not unusual for someone with respiratory problems to go to the hospital. This woman's symptoms were so severe, however, that the hospital decided to test for avian influenza, or bird flu. By the time the test results came back, she was dead.

This Indonesian woman tested positive for bird flu. Within four days of feeling ill, she had died. This was Indonesia's second confirmed human death from bird flu. It was confirmed that she had contracted the virus from fertilizer made from chicken waste. Her nephew also tested positive for the flu but recovered. There have been more than 200 confirmed cases of bird flu in humans and approximately half have died; therefore, bird flu has a 50% mortality rate for humans. Bird flu first appeared in 1997 and for eight years remained soley in China and southeast Asia. Suddenly, in 2005, its range expanded to Mongolia and Siberia, all the while maintaining its lethal nature. Bird flu has also been reported in poultry in Turkey and Romania and among migratory wild birds in Croatia.

AVIAN INFLUENZA VIRUS

The avian influenza or bird flu virus (*avian* means "bird') occurs naturally in birds. Wild birds have these viruses in their intestines and do not get sick from them, just as people have microbes in their intestines that do not make them sick.

Domesticated birds, however, are not resistant to influenza viruses. Avian influenza can be transmitted from wild to domestic birds through contact with an infected bird's secretions. Although the virus needs a host to live in, it can live for a period of time outside a body if it is in a bodily secretion. Domesticated birds, such as chickens, ducks, and turkeys, can catch bird flu and die.

Domestic birds can develop two different forms of bird flu. In the low-pathogenic form, the only noticeable difference in the bird is ruffled feathers and a drop in egg production. The highly pathogenic bird flu spreads rapidly through a flock, and the virus affects internal organs. The mortality rate for the highly pathogenic form of this disease is 90 to 100% in birds, usually within 48 hours.

Currently, bird flu is believed to be transmitted from birds to humans, rather than from humans to humans. People that get bird flu likely came into contact with an infected bird (Figure 8.1). This is significant because it is much easier to control the spread of a flu virus when the source is strictly birds instead of other people.

Even though there is a relatively low risk of someone catching bird flu, the risk still exists. The risk relies on the ability of viruses to change, making them able to invade the human body and do damage.

To understand the avian influenza virus, it is important to understand subtypes. A subtype is a virus that is similar to but different from the main type of virus. In the case of the influenza virus, subtypes are classified by changes in proteins on the surface of the influenza A virus. The proteins that change in the influenza A virus are hemagglutinin (HA) and neuraminidase (NA). There are 16 known HA subtypes and

Figure 8.1 This family in Nigeria is living in close daily contact with chickens. This lifestyle, typical of developing countries, facilitates transmission of the avian flu from bird to human. The fatality rate for people who contract bird flu is approximately 50%.

9 known NA subtypes of influenza A viruses. The different combinations of HA and NA proteins are staggering, and each combination represents a new subtype. This explains how keeping up with changes in the virus presents a challenge for providing effective vaccines.

Avian influenza viruses refer to influenza A viruses that are found mainly in birds, although infections in humans can occur. **Human influenza viruses** are the subtypes that spread among humans and are specifically referred to by their proteins: H1N1, H1N2, and H3N2. As mentioned, scientists believe that the subtypes of the human flu viruses are bird flu viruses that adapted to infect humans. The bird flu that is discussed in this chapter is also called the H5N1 virus.

HOST HOPPING

Live-animal markets are common in Asia and bring animals and people together to trade microbes. Most of the time, a virus is specific to one host, particularly when it is relatively new. A virus that is originally restricted to one host can mutate, however, and be able to infect another host directly and easily. When a virus that was restricted to an animal as host mutates, it often expands its host range to include humans. The virus usually becomes more harmful because humans have not evolved with this virus. It poses a new threat to the human body. Immunity would not exist even if the individual had gotten other flu vaccines. This is the concern with bird flu.

All influenza viruses got their start in birds, and the majority stay there. A handful of flu viruses, however, adapt to the point that they can infect people.[25] Of the bird flu viruses that have crossed the species barrier to infect humans, H5N1 has caused the largest number of detected cases of severe disease and death in humans.

All organisms adapt or mutate over time to make them better competitors. So, these microbes are doing what all other living organisms do. In most cases, we do not even notice, but if a newly emerged pathogen is deadly, we will feel the brunt of this evolutionary adaptation.

Viruses that have been known to infect people can present an additional problem if they mutate enough for fresh outbreaks to occur. A new vaccine is needed to fight the new virus strain. Generally, a person will have partial immunity to the new strain because she has been exposed to a similar virus.

Occasionally though, a strain that infects only birds will cross over more-or-less intact into humans. Because this new strain is so different from garden-variety flu viruses, few people are immune.[26]

Flu viruses are highly contagious. Most experts agree that the bird flu will become an epidemic; it is just a matter of time. The timing is most closely linked to when the virus can spread from human to human.

Infection, Spread, and Symptoms of Bird Flu

When a person sneezes, small droplets of water containing millions of viruses are released into the air with great force. Although a virus needs a host to survive long term, the virus can live for up to two days on a cold doorknob. As a result, most cases of the flu spread easily. In the case of bird flu, it is unclear whether there is any spread from one person to another. Some scientists believe that all infected people got it from close contact with an infected bird. Most scientists believe the virus will mutate and then will be able to spread from person to person. As of August 2006, bird flu does not exist in the United States. However, all it would take for this virus to become a threat here would be for an infected person to board a plane to the United States.

Another possibility for spread is for migratory birds with the disease to fly from an Asian country to Europe, the United States, or any other country without the disease. Removing sick birds from flocks of poultry has been effective in controlling the disease thus far but this method would not be possible for migratory birds.

Bird flu viruses have infected people before, but this particular strain, the H5N1 strain (Figure 8.2), is one of the most lethal, causing the largest number of cases of severe disease and death.[27] Coughing, sneezing, and fever are the initial symptoms. The virus ravages the lungs of those it infects within days, often leading to pneumonia and possibly to multiorgan failure. Currently there are no easy remedies. Often the patient is only offered oxygen to help them breathe. Too much oxygen, however, can burst a patient's lungs. A person's immune system will also attempt to fight the virus.

Cooking destroys the virus, so there is no risk from cooked poultry and eggs. Those who handle the manufacturing of

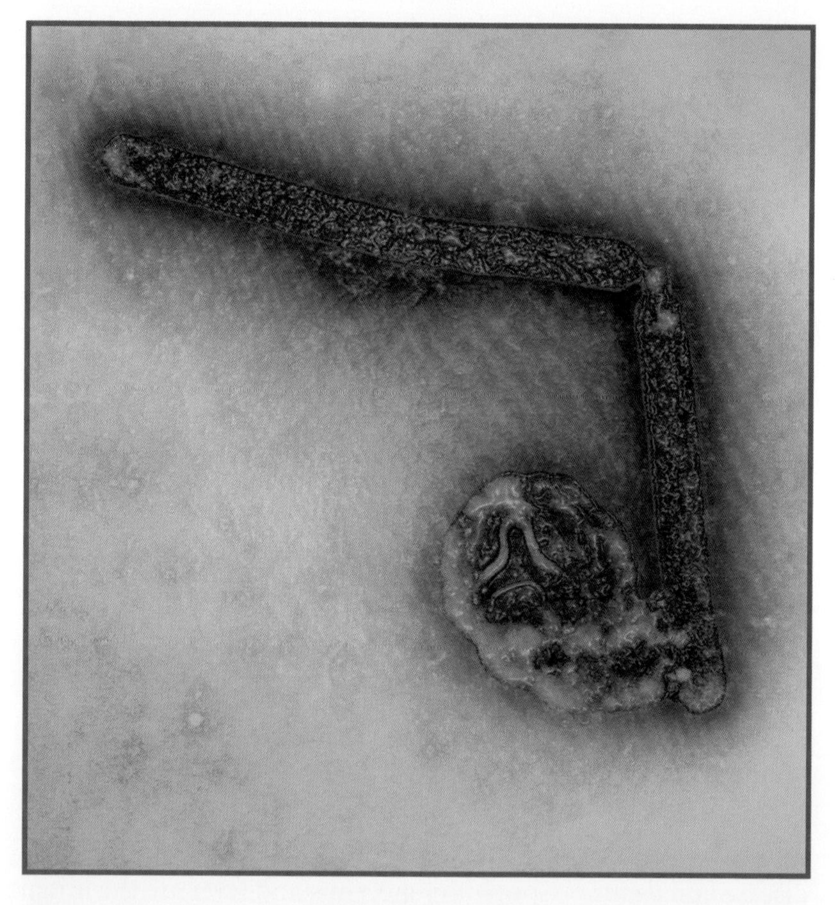

Figure 8.2 This is the H5N1 bird flu virus, as seen through a high-powered microscope. This particular strain of flu virus is one of the most lethal. The H5N1 virus invades the body and damages the lungs quickly, often leading to pneumonia and possible failure of the body's organs.

these foods would have reason to be concerned however. The European Union has already announced that it is banning chicken imports from Turkey after some birds there died of bird flu.

If an epidemic were to break out, keeping a distance from other human beings would be one way to avoid infection. Overcrowded places, such as buses, subways, and planes, would be a haven for the spread of this disease. Some bird flu experts have admitted to stockpiling a few months of food in case of an outbreak without an available vaccine. SARS was much less contagious than the bird flu virus, but during that 2003 outbreak in Hong Kong, people wore surgical masks and carried tissues to touch public surfaces such as doorknobs. This might be the case if bird flu is spread in the United States and can pass from human to human. It is important to note, however, that as of now, bird flu can only be caught from birds.

Treatment for Bird Flu

When a new vaccine against an influenza virus is being developed, scientists select the virus strain that will fight the enemy virus. The flu strain is carefully injected into the embryos of chicken eggs, which become tiny incubators for the virus. The virus is then killed with detergents, extracted, and bottled in vials.[28] This is how the vaccine for bird flu is being developed.

A vaccine seems like a simple solution to disease, but there are many shortcomings in the process. The process of making the flu vaccine has changed little since the 1940s. The process is slow and can take up to eight months. In the time frame of a bird flu pandemic, millions would die while waiting for the vaccine to become available to the general public. An influenza or flu pandemic is a global outbreak of disease that occurs when a new influenza virus emerges in the human population, causes serious illness, and spreads easily from person to person worldwide.

Another major concern is that the virus mutates and therefore would require another vaccine. One of the most crucial aspects of this process is understanding the strain of influenza that causes the illness. Many experts fear that, by the time the vaccine has been developed to fight bird flu, the virus will already have changed enough to render any current vaccine useless.

Viruses mutate naturally, but there are circumstances where a mutation might be more likely. One danger would be if a person working with poultry became infected with bird flu and then also caught a typical winter flu virus. The two viruses would have an opportunity to mix and possibly genetically alter to create a new strain. This new strain of virus would not necessarily respond to current vaccines or antiviral drugs.

The third major concern about treating a pandemic is simply not having enough vaccine to go around. As the concern over a pandemic grows, countries are trying to take steps to prevent the kind of devastation that a pandemic will cause. Governments are looking to develop and stockpile vaccines and antiviral drugs. Three private companies currently make vaccines. They are accustomed to producing enough vaccines to handle a typical year of flu, but not a pandemic. The government of Great Britain began stockpiling Tamiflu in 2006 for a quarter of their own population. The U.S. government is stockpiling enough for roughly 2 percent of the U.S. population. Critics argue that this is hardly enough to prevent widespread disease.

One way to ensure that there are enough vaccines to go around is to reduce the dose without lessening the effectiveness. Scientists recently discovered a way to do this: the vaccine is administered just under the skin instead of into the muscle. A vaccine injected just under the skin does not require as much material to provoke an immune response.

Antiviral drugs

Antiviral drugs, more simply referred to as antivirals, are also being developed to control bird flu. **Antivirals** aim to disable the destructiveness of the virus. A virus needs a host to live. To reproduce, the virus must attach to the host's cell and then "uncoat" it to gain entry. The virus then tricks the host's cell into making copies of the virus. Antivirals disable a virus either by hindering its entrance into the cell or by blocking the release of the virus's genetic material, which tricks the invaded cell to make more viral particles. Most flu antivirals work by disabling the release of the virus's genetic material, which enables it to reproduce. Unlike vaccines, antivirals do not develop immunity in a host; antivirals protect the host by disabling the destructiveness of the virus.

One main advantage of antivirals is that they can be used for any strain of the flu. This resolves the dilemma of what to do when the virus changes genetically. One shortcoming of antivirals is that their effectiveness varies depending on the strength of the virus. Fludase and Tamiflu are two antivirals currently developed for bird flu.

Fludase works by preventing the initial entrance of the virus into the cell. Scientists believe that, by blocking entrance into a cell, resistance to an antiviral will not develop. Tamiflu is another antiviral that must be taken within 48 hours of the onset of symptoms and requires a prescription. One tablet is taken in the morning and the other is taken at night for five days at a cost of roughly $80 to $90. Tamiflu is meant to lessen the severity of the illness, which would save lives. If someone knew that they came in contact with an infected bird, they could take Tamiflu as a preventative measure. Yet, even with a limited use of Tamiflu, there has already been a case of resistance. A girl in Vietnam had bird flu, was given Tamiflu, and did not respond. Researchers discovered that she had a resistant strain of the virus.

Although these antivirals have been developed, there have been no clinical trials on humans. Specifically, their effect on children younger than 12 is unknown. Roche, a private pharmaceutical company that is the only supplier of Tamiflu at present, is negotiating with other companies to manufacture its product. In Taiwan, officials have announced production of their own generic version of Tamiflu, which they claim is 99% similar to the original. This flurry of production raises concern about quality control.

People concerned about bird flu and recognizing the potential shortfall if an outbreak were to occur, are attempting to stockpile their own Tamiflu. The *Washington Post* reported that 1.7 million prescriptions were filled in the United States in the first eight months of 2005, three times the rate in 2004.[29] Consumers should be careful of offers of Tamiflu over the Internet as there has been speculation about counterfeit Tamiflu.

Policy and Public Health

Disease is a public health concern. Scientists have likened the effects of pandemic bird flu to a breakdown in societal order, as was seen in New Orleans after Hurricane Katrina. In a pandemic, however, all major cities would be affected. In the grimmest of scenarios, store shelves would be empty, mail delivery would stop, and gas would be in short supply as thousands or millions died.

Fighting pathogens is costly. The bird flu is a good example of how expensive it is to protect people and animals from these microbes. At the end of 2005, the World Bank estimated that it would cost a billion dollars to control an outbreak or pandemic of bird flu. More than 100 nations agreed to an international effort, and each participating country is pledging millions of dollars to the effort.

The U.S. government appropriated $1.2 billion to support drug companies' efforts to develop vaccines and stockpile them. As mentioned, however, this effort would only provide enough vaccines for 2% of the American population. Most European countries have a goal of stockpiling enough vaccines for 25% of their populations.

The other option is to do nothing and let the flu run its course. This is a dismal alternative considering that, even with control efforts currently in place, the Food and Agriculture Organization estimates the economic impact has been more than $10 billion.[30] At the time of this writing, experts estimate that the chances of a pandemic of bird flu are nearly 100%.

Governments recognize the importance of working together to fight this pandemic. Microbes do not recognize national boundaries, so an outbreak of bird flu in Vietnam is of concern to all nations. Bird deaths in Turkey and Romania have already shown that the virus is moving westward. Even if the virus can only be spread by birds and not humans, the devastation to Western agriculture alone would be crippling.

All nations need to work together to control this flu, yet many countries simply cannot afford antivirals or other technologies to properly handle this health concern. Signs of a coordinated, international effort do exist. On September 14, 2005, the U.S. government announced an International Partnership on Avian and Pandemic Influenza to improve international surveillance and response. The European Union is considering offering funding to Asian countries to combat bird flu.[31] Controlling bird flu at the source is the most cost-effective option.

PREVIOUS FLU PANDEMICS

Flu pandemics generally occur every 30 years. The predictions of what would happen in the case of pandemic bird flu are drawn from the history of pandemics. The last time a flu virus reached pandemic levels was in 1968. It killed 34,000 Americans.

The 1918 flu pandemic was more noteworthy with regard to viral strength and subsequent deaths. The 1918 pandemic killed 675,000 Americans, and the U.S. population was only 100 million.[32] Today, the U.S. population is more than double that figure. Worldwide, 50 to 100 million have died from this flu. There were more deaths within 24 weeks than AIDS has killed in 24 years.[33]

People who lived through the 1918 flu pandemic still speak about the horrific suffering. People bled from the eyes and ears, and their bodies turned a color so dark that it was difficult to determine race. Many sick people died because the healthy were afraid to come into contact with them. Survivors recall that funerals would be held with hardly anyone in the church because nearly everyone else was already dead. Waking up in the morning would often bring the news of someone else in the family or neighborhood who had died during the night.

The U.S. government was blamed for mishandling the 1918 flu pandemic. The devastating effects were minimized at the outset, which created a bigger problem because people did not know how to respond.

PREPARING FOR A PANDEMIC

Private companies are already responding to public concerns. A company called Respirators Inc. is expanding its line of respirators in response to the concern over bird flu. In addition, this company is selling antiseptic hand sprays, gels, and wipes as well as latex gloves. These products are not currently endorsed by the Centers for Disease Control (CDC) because this flu does not yet pass from human to human. These products would, however, reduce the risk of transmission in the event that the virus became capable of human-to-human transmission.

Despite the fact that bird flu cannot be transmitted from human to human, Respirators Inc. reported orders in 2005 that ranged from 20 to 500 respirators. Recent publications have mentioned respirator companies arming their traveling executives with masks, gloves, and Tamiflu.

People with a weakened immune system, such as the elderly, should be most concerned about a pandemic. Unless there is a radical change in the production of vaccines or antivirals, many people will be relying on the strength of their immune system to fight bird flu.

9 Managing the Invasive Microbes

Invasive microbes threaten the biodiversity of our landscape by killing native species. **Biological diversity,** or **biodiversity,** is the richness of life-forms in nature. The more diversified the life-forms are in an ecosystem, the more resilient and stable that ecosystem will be. Invasive microbes make our ecosystems less resilient by reducing biodiversity.

Invasive microbes are able to thrive because they have a competitive edge. Instead of microbes that have evolved with native animals and plants, international travel is introducing exotic microbes against which native organisms have never had to develop defenses. The fact that they are newly introduced to an environment is the very reason they are able to have the competitive edge.

The destructiveness of a newly introduced microbe is well proven in whirling disease in fish. The parasite that causes whirling disease was introduced from Europe. European fish, such as the brown trout are not as severely affected by this parasite because the parasite and brown trout evolved together and developed traits to handle their adversary. Fish native to the United States did not evolve with the whirling disease parasite; so native fish are more susceptible to the parasite that causes this disease.

INVASIVES IN BALLAST WATER

Open sea exchange has been proposed for controlling invasive species, but it is not effective in preventing the introduction of invasive microbes. The nozzle that sucks ballast water from each tank and pumps it from the ship hangs far into the tank, but above the actual floor. A few inches of water always remains beneath the nozzle. Sediment that settles out of ballast water also stays behind.[34] Because the water beneath the nozzle is not removed, even ships that are considered free of ballast water usually contain dozens of species foreign to the port they are going to. To control microbes in ballast water, a biocide is needed, something that actually kills the organism and breaks down quickly.

A biocide, such as vitamin K, would resolve the problem of even a small amount of water left in the tank. Vitamin K provides a 90 to 95% kill rate for organisms, which still leaves a concern for the remaining percentage. Vitamin K breaks down in the water after a day, so there is no harm to the aquatic life in the port when it is released.

If the ballast water needs to be discharged in a short amount of time, ozone and nitrogen gases are an alternative. These gases are already used to control bacteria in pools and aquariums, and would work well on ballast water. Ozone gas effectively kills the larvae and eggs of fish and crustaceans, but does not work very well on adult organisms. This is not a major issue, however, because the eggs and larvae are more of a concern in ballast water than the mature animals. A major advantage of the ozone is that it is very reactive so it breaks down quickly in water.

Nitrogen gas pumped into the tank also works well without using chemicals. When nitrogen is bubbled through the tank, the available oxygen, which the organisms need to live, is lowered. Scientists have found that the organisms die within three days. The nitrogen eventually bubbles out, and the oxygen concentration returns to normal so the ballast water will not injure aquatic life when it is dumped in the port.

It might be easier to persuade crews to use this method; it has an added benefit of lowering the levels of oxygen and thereby reducing rusting in the tank. The disadvantage of using nitrogen gas is that it is more expensive.

No perfect solution exists for stopping the introduction of exotic microbes in ballast water, although scientists agree that using at least one method is better than doing nothing. The biggest challenge is in regulating an international community. Forcing ships to kill organisms in the ballast water is difficult to enforce, but not impossible.

Ballast Water Treatment

No single management technique has proven successful in killing or removing all organisms in ballast water. In open sea exchange, a ship would empty the ballast water in the open sea, such as the Atlantic Ocean instead of the Great Lakes, and then fill the ballast tank with ocean water. This method works because coastal organisms are unlikely to be able to survive in the open sea. The option has the appeal of being easily monitored; a simple salinity test would be able to detect whether the ballast water is seawater or fresh. Open sea exchange is not safe, however, when the seas are stormy or rough. Another limitation is that sediment and residual water are difficult to remove from the ballast tank.

Biocides can also be used to kill organisms in ballast water, but one concern is the health of the crew handling these chemicals. Another concern is the potential of corroding the ballast tank. Ballast water can be heated to temperatures between 95^{0}F to 113^{0}F to kill larger organisms such as fish, but not microorganisms. No easy solution to this major pathway of biological invasions exists. Until recently, treating ballast water was not a high priority. Research continues, however, for techniques to treat ballast water without jeopardizing the safety of crew members.

Hull Fouling

Sea ballast water accidentally carries organisms inside the ship, whereas hull fouling carries organisms outside the ship where they attach to the hull. Open sea exchange will not reduce hull fouling. Organisms such as barnacles, mussels, sea squirts, sponges, and algae are able to attach to the hull and be transported long distances (Figure 9.1). Aquatic animals can be vectors (carriers) for microbes. Once they arrive in a new port, they can create new exotic populations by releasing their larvae or attaching to another structure in the port. Hull fouling is easily solved by building metal hulls and using antifouling paints to prevent organisms from attaching to the hull. As shipping increases, it becomes more vitally important to incorporate preventative measures, such as antifouling paint, to prevent an exotic species explosion.

Figure 9.1 Zebra mussels are an invasive species that has been accidentally introduced by attaching to the hulls of ships. In this photo, the tiny zebra mussels attach to a native mussel. Zebra mussels disrupt native aquatic life and clog water supply pipes.

Onshore Treatment

Onshore treatment removes ballast water from a ship when it enters port to be treated, or the ballast water is treated prior to being loaded onto a ship. The water can be treated in facilities dedicated to ballast water or in facilities for treatment of wastewater. Water can also be stored and recycled for use as ballast by other ships.

Although it sounds tedious to treat water, it costs about the same as other treatment methods. Onshore treatment has the advantage of not putting crew members' lives at risk by exchanging water on the open sea or requiring them to have hazardous materials onboard. Cheap initial treatments, such as sedimentation, can be used onshore rather than on the ship because of the availability of space onshore. Sedimentation may be capable of removing many resistant life stages, such as cysts, which are capsules that contain parasites, and spores as well as organic and inorganic suspended sediment, making subsequent treatment (UV or biocides) cheaper and more efficient.

EDUCATION OF THE PUBLIC

Action occurs through an educated public. Organizations are formed at all levels to control invasive species. The nonprofit environmental group, The Nature Conservancy, looks for invasive species on preserved lands and creates management plans to control or eradicate the exotics that they find. On their Web site, The Nature Conservancy educates members and others with their America's Least Wanted: Alien Species Invasions of U.S. Ecosystems (http://www.natureserve.org/publications/leastwanted).

The U.S. government formed the National Invasive Species Council to coordinate national efforts. A large part of these efforts also consists of educating the public. One way they do this is by providing species profiles, along with information about local laws and regulations (http://www.invasivespecies.gov).

The World Conservation Union's Invasive Species Specialist Group enlists the help of scientists and policy experts from around the world to collect and distribute information about where invasive species are living and growing their populations.

A NATIONAL PLAN TO CONTROL INVASIVE MICROBES

Strategies to control invasive microbes exist on the small and large scale. All organizations that manage land or commerce need to implement a plan to prevent the introduction of invasive microbes as well as the spread of already introduced ones.

Executive Order 13112, issued in 2001, is the large-scale plan for handling invasive species. This order created the Invasive Species Management Plan that is the U.S. government's plan of action.

The executive order is just a start and lacks some specifics. The plan does not address the intentional introduction of invasive species as a security threat. The plan also is not being implemented quickly enough. The longer it takes to put programs into place, the more acres of land and native species we lose daily to invasive species.

Another deficiency of the Invasive Species Management Plan is that it does not address invasive pathogens. Although invasive pathogens are included in the definition of invasive microbes, these disease-causing invasive microbes need to be handled with a quicker response and different strategies to reduce the number of people who get sick and possibly die from them.

Opinions vary on exactly how destructive invasive species are, but most naturalists embrace the notion that all species have value and to lose them haphazardly by shuffling invaders around the globe is foolish. In some cases, we may be losing species that we have yet to discover, and that is like throwing away a good book before it has been read. Some of the great challenges to stopping the spread of invasive species are in changing people's perceptions and increasing awareness about the issue.

NOTES

1. Anne Simon, *The Real Science Behind the X-Files.* New York: Simon and Schuster, 1999, p. 24.
2. Tony Hart, *Microterrors.* London: Axis Publishing, 2004, p. 10.
3. American Society for Microbiology, "Why Are Microbes Such an Evolutionary Success Story?" 1999. Available online at: http://www.microbe.org/microbes/success.asp.
4. Anne Simon, *The Real Science Behind the X-Files.* New York: Simon and Schuster, 1999, p. 27.
5. Tony Hart, *Microterrors.* London: Axis Publishing, 2004, p. 10.
6. Ibid., p. 11.
7. Ibid., p. 54.
8. Ibid., p. 55.
9. Ibid., p. 6.
10. "Q&A: MRSA 'superbug'" *BBC News.* April 16, 2006. Available online at: http://news.bbc.co.uk/go/pr/fr/-/1/hi/health/2572841.stm.
11. *The Economist.* "The Usual Suspects," (November 9, 2005), p. 84.
12. Ibid.
13. A.W. Crosby, Jr., *The Columbian Exchange.* Westport, Conn.: Greenwood Press, 1972, p. 219.
14. Mary E. Wilson, "Travel and the Emergence of Infectious Diseases," *Emerging Infectious Diseases* 1, no.2 (1995): p. 40.
15. Ibid., p. 39.
16. Ibid., p. 40.
17. *The Economist,* "The Usual Suspects," (November 9, 2005), p. 85.
18. G.N. Agrios, *Plant Pathology.* New York: Academic Press, Fourth Edition, 1997.
19. Ellen Michaels Goheen, "Detecting, surveying, and monitoring *Phytophthora ramorum* in forest ecosystems." Sudden Oak Death online symposium, p. 30. Available online at: http://www.apsnet.org/online/SOD.
20. Ibid.
21. Ibid., p. 31.
22. D. Michael Benson, Cultural practices and host resistance: Two IPM strategies for control of *Phytophthora ramorum* in nurseries. Sudden Oak Death online symposium. Available online at: http://www.apsnet.org/online/SOD/pdf/benson.pdf.
23. Jason Stein, "Is *Ralstonia* A Weapon Terrorists Would Use?" *Wisconsin State Journal.* (July 6, 2003) Available online at: http://www.highbeam.com.
24. Adrian Higgins, "Geranium Blight Prompts Worry." *The Cincinnati Post.* (January 17, 2004.) Available online at: http://www.highbeam.com.
25. Christine Gorman, "How Scared Should We Be?" *Time.* (October 17, 2005), p. 32.

26. Ibid.
27. Jeremy Laurance, "Bird Flu: Everything You Need to Know." *The Independent.* London: Independent Newspapers, October 18, 2005. Available online at: http://www.highbeam.com.
28. Michael Rosenwald, "New Bird Flu Vaccine Might Not Work in Time." *Discover.* (January 2006), p. 44.
29. Jerry Adler, "The Fight Against the Flu." *Newsweek.* (October 31, 2005) Available online at: http://www.highbeam.com.
30. *The Economist,* "The Usual Suspects," (November 9, 2005), p. 84.
31. Emma Dorey, "Ineffective flu strategy warning." *Chemistry and Industry.* (October 3, 2005)
32. John M. Barry, "Lessons from the 1918 Flu," *Time.* (October 17, 2005), p. 96.
33. Ibid.
34. Ben Harder,"Stemming the tide; killer technologies target invading stowaways." *Science News.* (April 13, 2002).

GLOSSARY

Antagonistic Relationship between competing microbes that maintains control over an invasive microbe's growth.

Antivirals Drugs or agents that aim to disable the destructiveness of a virus.

Asexual reproduction Reproduction by dividing in two.

Avian influenza viruses Influenza A viruses that are found mainly in birds.

Ballast water Water that is pumped into a ship to make the vessel heavier and therefore more stable in transit along coasts and on the open seas.

Biological control The method of introducing an invasive species' natural predator to control an invasive species.

Biological diversity (biodiversity) The richness of life-forms in nature.

Biotechnology The use of living things to make products useful to humans.

Botany The study of plants.

Cankers Places on a tree, such as the bark, where the tissue has died and eventually rots.

Chlorophyll The green pigment that enables a plant to use sunlight and carbon dioxide to make food to support its structure and function.

Chlorosis Yellowing of the leaves due to the loss of chlorophyll.

Coevolution Process in which two species of organisms coexist and evolve in response to a similar environment.

Disease vector An organism that by itself may not cause disease, but the microbes on it do cause disease.

Emerging pathogen A pathogen that has emerged relatively recently, such as SARS or HIV; 13% of all human-related pathogens are considered to be emerging pathogens.

Endosymbiosis Condition in which a microbe lives in harmony within the body of another organism.

Entomology The study of insects.

Eukaryotes Life-forms that have cells with a nucleus.

Genetically Modified Organisms (GMOs) Organisms with intentionally modified genetic material.

Host-specific Referring to an organism with very specific feeding habits; this organism generally consumes only one species, specifically the organism that is trying to be suppressed.

Human influenza viruses The subtypes that spread among humans and are specifically referred to by their proteins: H1N1, H1N2, and H3N2.

Hyphae Microscopic threadlike filaments that branch out and feed the fungus.

Integrated pest management (IPM) A strategy to control plants, animals, and microorganisms by using more than one of these methods: chemical, biological, cultural, mechanical.

Invasive species Animals, plants, and microbes that invade ecosystems beyond their historic range.

Mutually beneficial A relationship in which both host and parasite benefit.

Mycelium Dense mold.

Mycorrhizal fungi Beneficial fungi that live in and around plant roots that get minerals and water out of the soil, which are then made available to the plant.

Myxospore Mature spore.

Nitrogen-fixing bacteria Bacteria that take nitrogen from the air and change it into an available form, or "fix" it, to make it accessible to plants. The plants in turn create nodules on their roots in which bacteria can live.

Nucleus A central compartment in a cell.

Pandemic A global outbreak of disease.

Pathogens Disease-causing microbes.

Photosynthesis Process by which a plant feeds itself by using chlorophyll and sunlight to make sugars.

Plant pathology The study of plant disease.

Plasmodia The vegetative body of fungi; plasmodia are amorphous, jelly-like slime molds.

Prokaryotes Organisms with cells that do not have their genetic material enclosed in a nucleus.

Spores The reproductive element of simple organisms, such as protozoa, fungi, and nonflowering plants such as ferns.

Threshold level A specified population size of microbes required to control the pathogens.

Triactinomyxon (TAM) A highly infective parasite of fish that metamorphoses from a spore.

Virosome vaccine Type of vaccine consisting of a virus that is unable to replicate.

Viruses Smallest microbe that attaches to a host cell to reproduce.

Zoonoses Diseases that come from animals.

BIBLIOGRPAHY

Adler, Jerry. "The Fight Against the Flu." *Newsweek.* (October 31, 2005). Available online at: http://www.highbeam.com.

Agrios, G.N. *Plant Pathology.* New York: Academic Press, Fourth Edition. 1997.

American Society for Microbiology. "Why Are Microbes Such an Evolutionary Success Story?" 1999. Available online at: http://www.microbe.org/microbes/success.asp.

APHIS Services, USDA. "Detection of *Ralstonia Solanacearum* Race 3 Biovar in the United States." March 2003. Available online at: http://www.aphis.usda.gov/lpa/pubs/fsheet_faq_notice/fs_phralstonia.html.

APHIS Services, USDA. "Exotic Newcastle Disease." January 2003. Available online at: http://www.aphis.usda.gov/lpa/pubs/fsheet_faq_notice/fs_ahend.html.

Barry, John M. "Lessons from the 1918 Flu." *Time.* (October 17, 2005).

Benson, D. Michael. "Cultural practices and host resistance: Two IPM strategies for control of *Phytophthora ramorum* in nurseries." Sudden Oak Death online symposium. Available online at: http://www.apsnet.org/online/SOD.

Boselovic, Len. "Scientists find steel pollutant-eating bacteria." *The Augusta Chronicle.* (February 8, 2002). Available online at: http://www.highbeam.com.

Breidahl, Harry. *Extremophiles: Life in Extreme Environments.* Langhorne, Pa: Chelsea House Publishers. 2002.

Colorado Division of Wildlife. "Whirling Disease and Colorado's Trout." Available online at: http://wildlife.state.co.us/fishing/whirling.asp.

Crosby, A.W., Jr. *The Columbian Exchange.* Westport, Conn.: Greenwood Press. 1972.

Department of Health and Human Services, Centers for Disease Control and Prevention (CDC). Division of Bacterial and Mycotic Diseases, "Legionellosis: Legionnaire's Disease (LD) and Pontiac Fever. Available online at: http://www.cdc.gov/ncidod/dbmd/diseaseinfo/legionellosis_g.htm.

Department of Health and Human Services, Centers for Disease Control and Prevention (CDC). "Key Facts About Avian Influenza (Bird Flu) and Avian Influenza A (H5N1) Virus." (November 25, 2005). Available online at: http:// www.cdc.gov/flu.

Dorey, Emma. Society of Chemical Industry. "Ineffective flu strategy warning." *Chemistry and Industry.* (October 3, 2005). Available online at: http://www.highbeam.com.

Dusenbery, David B. *Life at Small Scale: The Behavior of Microbes.* New York: Scientific American Library. 1996.

Garbelotto, Matteo. "Molecular diagnostics of *Phytophthora ramorum*, casual agent of Sudden Oak Death," Sudden Oak Death online symposium. Available online at: http://www.apsnet.org/online/SOD.

Gardner, Den. "Learning the lessons. Asian soybean rust, farmers of Brazil." *Apply.* Primedia Business Magazines and Media. (April 1, 2005). Available online at: http://www.highbeam.com.

Goheen, Ellen Michaels. "Detecting, surveying, and monitoring *Phytophthora ramorum* in forest ecosystems." Sudden Oak Death online symposium. Available online at: http://www.apsnet.org/online/SOD.

Gorman, Christine. "How Scared Should We Be?" *Time.* (October 17, 2005).

Groves, Bob. "Migratory birds could unleash avian flu here," *The Record.* (October 28, 2005). Available online at: http://www.highbeam.com.

Harder, Ben. "Stemming the tide; killer technologies target invading stowaways." *Science News.* (April 13, 2002). Available online at: http://www.highbeam.com.

Hart, Tony. *Microterrors.* London: Axis Publishing. 2004.

Henderson, Pam. "Meet Your Hosts: Asian Soybean Rust," *Farm Journal.* (January 16, 2005). Available online at: http://www.highbeam.com.

_________. "Soybean Rust Takes Root," *Farm Journal.* (January 4, 2005). Available online at: http://www.highbeam.com.

Higgins, Adrian. "Geranium Blight Prompts Worry." *The Cincinnati Post.* (January 17, 2004). Available online at: http://www.highbeam.com.

Kuchment, Anna. "Trapping the Superbugs; Antibiotics are losing their punch as microbes learn to resist them. Can we stop the new killers?" *Newsweek.* (December 6, 2004.) Available online at: http://www.highbeam.com.

Laurance, Jeremy. "Bird Flu: Everything You Need to Know." *The Independent.* (October 18, 2005). Available online at: http://www.highbeam.com.

Lipp, Linda. "Indiana farmers take action to minimize damage from soybean-rust fungus," *The News-Sentinel.* (January 4, 2005). Available online at: http://www.highbeam.com.

Mayse, James. "Soybean rust spores found in Kentucky, Tennesee." *Messenger-Inquirer.* Ownesboro: Kentucky. (July 8, 2005). Available online at: http://www.highbeam.com.

Momol, Timur, Tom Kucharek, and Hank Dankers. "Fungal and Bacterial Disease Diagnoses for Distance Diagnostic and Identification System (DDIS)," University of Florida, IFAS Extension. Available online at: http://edis.ifas.ufl.edu/DDIS2.

National Public Radio (NPR). "Profile: Exotic Newcastle Disease, a lethal avian virus spreading through California's poultry population." *All Things Considered.* (Aired on January 7, 2003).

Parris, Thomas. "Hunting Down Invasive Species," *Environment.* (December 1, 2002). Available online at: http://www.highbeam.com.

Pradhanang, Prakash M., Timur Momol, Hank Dankers, Esen A. Momol. "First Report of Southern Wilt Caused by *Ralstonia solanacearum* on Geranium in Florida," *Plant Management Network.* (June 11, 2002) Available online at: http://www.plantmanagementnetwork.org/pub/php/brief/geranium.

Pratt, Robert J. "Invasive threats to American homeland." *Parameters.* (March 22, 2004). Available online at: http://www.highbcam.com.

Rosenwald, Michael. "New Bird Flu Vaccine Might Not Work in Time." *Discover.* (January 2006). Available online at http://www.discover.com.

Rouquet P., J.M. Froment, M Bermejo, A. Kilbourne, W. Karesh, P. Reed , et al. "Wild animal mortality monitoring and human Ebola outbreaks, Gabon and Republic of Congo, 2001–2003. Emerging Infectious Diseases." (February 2005) Available online at: http://www.cdc.gov/ncidod/EID/vol11no02/04-0533.htm.

Simon, Anne. *The Real Science Behind the X-Files.* New York: Simon and Schuster. 1999.

Smith, Michael. "Bacteria Can 'Farm' Plants," *United Press International.* (May 26, 2002). Available online at: http://www.highbeam.com.

Soares, Christine. "Getting Serious About Flu," *Scientific American.* (December 2005).

Stein, Jason. "Is *Ralstonia* A Weapon Terrorists Would Use?" *Wisconsin State Journal.* (July 6, 2003). Available online at: http://www.highbeam.com.

Stein, Jason. "Floating Fungus Might Attack State Soybeans," *Wisconsin State Journal.* (December 4, 2004). Available online at: http://www.highbeam.com.

Systemic Botany and Mycology Lab, Agricultural Research Service, U.S. Department of Agriculture. "Asian Soybean Rust." Available online at: http://nt.ars-grin.gov/taxadescriptions/factsheets.

Todar, Kenneth. "*Vibrio cholerae* and Asiatic Cholera," *Todar's Online Textbook of Bacteriology*. University of Wisconsin-Madison 2005. Available online at: http://textbookofbacteriology.net/cholera.html.

Vinarsky, Cynthia. "Germy Geraniums Call for Quarantine in Ohio and Pennsylvania." *Vindicator.* Youngstown, Ohio: Tribune Business News. (April 5, 2003). Available online at: http://www.highbeam.com.

Werres, Sabine and Thomas Schroeder. "Nursery Detection," Sudden Oak Death online symposium. Available online at: http://www.apsnet.org/online/SOD.

Wilson, Mary E. "Travel and the Emergence of Infectious Diseases." *Emerging Infectious Diseases* 1, no. 2 (1995). (April–June 1995).

FURTHER READING

Baskin, Y. *A Plague of Rats and Rubbervines: The Growing Threat of Species Invasions.* Covelo, Calif.: Shearwater Books, 2003.

Cox, G. *Alien Species and Evolution: The Evolutionary Ecology of Exotic Plants, Animals, Microbes and Interacting Native Species.* Washington, D.C.: Island Press, 2004.

Forsyth, A. and K. Miyata. *Tropical Nature.* New York: Charles Scribner's Sons, 1984.

Mooney, H.A. *Invasive Species in a Changing World.* Washington, D.C.: Island Press, 2000.

Wilson, E.O. *The Diversity of Life.* Cambridge, Mass.: Belknap Press of Harvard University, 1992.

_________. *The Future of Life.* New York: Vintage Publishing, 2003.

Web Sites

Center for Disease Control and Prevention (CDC)

http://www.cdc.gov

Cornell University – Biological Control

http://www.nysaes.cornell.edu/ent/biocontrol/info/biocont.html

Eco-Pros Invasive Non-Native Species

http://www.eco-pros.com/invasive_non-native_species.htm

The Global Invasive Species Database

http://www.issg.org/database/welcome

Invasive Species in the Pacific, including Hawaii
http://www2.ctahr.hawaii.edu/adap2/hottopics/invasive_species.htm

The Nature Conservancy (TNC)-Invasive Species
http://www.nature.org/initiatives/invasivespecies

USDA Animal and Plant Health Inspection Service (APHIS)
http://www.aphis.usda.gov/ppq

USDA APHIS Pest Tracker
http://ceris.purdue.edu/napis/index.html

USDA National Invasive Species Information Center
http://www.invasivespeciesinfo.gov/

The Wildlife Society
http://www.wildlife.org/policy/index.cfm?tname=positionstatements&statement=ps14

PICTURE CREDITS

page:

8: Eye of Science/Photo Researchers, Inc.
12: Sara Wright/Agricultural Research Service, USDA
21: Scott Bauer/Agricultural Research Service, USDA
27: BURGER/Photo Researchers, Inc.
30: Phil Degginger/Dembinsky Photo Associates
34: Claudia Adams/Dembinsky Photo Associates
38: Stephen Ausmus/Agricultural Research Service, USDA
41: Chuck Young/U.S. Fish & Wildlife Service
45: Joseph O'Brien, USDA Forest Service, www.forestryimages.org
46: Scott Bauer/Agricultural Research Service, USDA
48: Lawrence Migdale/Photo Researchers, Inc.
51: Stephen Ausmus/USDA Agricultural Research Service, www.forestryimages.org
55: Scott Bauer/Agricultural Research Service, USDA
58: Thomas Ilias/Agricultural Research Service, USDA
59: Jean L. Williams-Woodward, The University of Georgia, www.forestryimages.org
66: AFP/Getty Images
69: Associated Press, AP
72: Reid Frederick, USDA Agricultural Research Service, www.forestryimages.org
79: Associated Press, AP
82: CDC/ScienceSource/Photo Researchers, Inc.
93: U.S. Fish and Wildlife Service

Cover: CDC / C. GOLDSMITH / J. KATZ / S. ZAKI / Photo Researchers, Inc.

INDEX

ABOUT THE AUTHOR

Suellen May writes for agricultural and environmental publications. She is a graduate of the University of Vermont (B.S.) and Colorado State University (M.S.). She has worked in the environmental field for 15 years, including invasive species management for Larimer County Open Lands in Colorado. She served as the Education Committee chairperson for the Colorado Weed Management Association. While living in Fort Collins, Colorado, she founded the Old Town Writers' Group. She lives with her son, Nate, in Bucks County, Pennsylvania. Readers can reach her at suellen0829@yahoo.com.